ROTARY DRILLING SERIES

Testing and Completing

Unit II, Lesson 5
Third Edition

By James Vaught

Published by

THE UNIVERSITY OF TEXAS AT AUSTIN
PETROLEUM EXTENSION SERVICE
Continuing Education
Austin, Texas

in cooperation with

INTERNATIONAL ASSOCIATION
OF DRILLING CONTRACTORS
Houston, Texas

2001

Library of Congress Cataloging-in-Publication Data

Vaught, James, 1935—
Testing and completing / by James Vaught. — 3rd ed.
p. cm.
"Unit II, lesson 5."
ISBN 0-88698-192-1 (alk. paper)
1. Oil well—Testing. I. Title.
TN871.V345 2001
622'.3382—dc21 2001002295
CIP

First Edition published 1968. Second Edition 1983.
Third Edition 2001. Third impression 2010.
Printed in the United States of America

Catalog no. 2.20530
ISBN 0-88698-192-1

Contents

Figures

Tables

Foreword

For many years, the Rotary Drilling Series has oriented new personnel and further assisted experienced hands in the rotary drilling industry. As the industry changes, so must the manuals in this series reflect those changes.

The revisions to both the text and illustrations are extensive. In addition, the layout has been 'modernized' to make the information easy to get. The study questions have been rewritten, and each major section has been summarized to provide a handy comprehension check for the reader.

PETEX wishes to thank industry reviewers—and our readers—for invaluable assistance in the revision of the Rotary Drilling Series.

Testing and Completing introduces rig crew members to well test procedures and to various completion methods operators use to finalize a well. Further, it covers reservoirs and reservoir characteristics to give readers a foundation in formation evaluation, formation testing, and completion design. Completion equipment is also addressed.

Although every effort was made to ensure accuracy, this manual is intended only as a training aid; thus, nothing in it should be construed as approval or disapproval of any specific practice or product.

Ron Baker

Acknowledgments

Sincere appreciation is extended to those who helped with the preparation of the third edition of *Testing and Completing*. The following companies are those that provided materials and answered questions:

Barite Rose Energy

B. P. McBride Engineering

Halliburton Energy Services

Schlumberger Well Services

Special thanks are extended to Annette Vaught for her tireless review of the text and invaluable comments and suggestions. This input was especially helpful in the preparation of text from a nontechnical standpoint.

Units of Measurement

Throughout the world, two systems of measurement dominate: the English system and the metric system. Today, the United States is almost the only country that employs the English system.

The English system uses the pound as the unit of weight, the foot as the unit of length, and the gallon as the unit of capacity. In the English system, for example, 1 foot equals 12 inches, 1 yard equals 36 inches, and 1 mile equals 5,280 feet or 1,760 yards.

The metric system uses the gram as the unit of weight, the metre as the unit of length, and the litre as the unit of capacity. In the metric system, for example, 1 metre equals 10 decimetres, 100 centimetres, or 1,000 millimetres. A kilometre equals 1,000 metres. The metric system, unlike the English system, uses a base of 10; thus, it is easy to convert from one unit to another. To convert from one unit to another in the English system, you must memorize or look up the values.

In the late 1970s, the Eleventh General Conference on Weights and Measures described and adopted the Système International (SI) d'Unités. Conference participants based the SI system on the metric system and designed it as an international standard of measurement.

The *Rotary Drilling Series* gives both English and SI units. And because the SI system employs the British spelling of many of the terms, the book follows those spelling rules as well. The unit of length, for example, is *metre*, not *meter*. (Note, however, that the unit of weight is *gram*, not *gramme*.)

To aid U.S. readers in making and understanding the conversion to the SI system, we include the following table.

English-Units-to-SI-Units Conversion Factors

Quantity or Property	English Units	Multiply English Units By	To Obtain These SI Units
Length, depth, or height	inches (in.)	25.4	millimetres (mm)
		2.54	centimetres (cm)
	feet (ft)	0.3048	metres (m)
	yards (yd)	0.9144	metres (m)
	miles (mi)	1609.344	metres (m)
		1.61	kilometres (km)
Hole and pipe diameters, bit size	inches (in.)	25.4	millimetres (mm)
Drilling rate	feet per hour (ft/h)	0.3048	metres per hour (m/h)
Weight on bit	pounds (lb)	0.445	decanewtons (dN)
Nozzle size	32nds of an inch	0.8	millimetres (mm)
Volume	barrels (bbl)	0.159	cubic metres (m^3)
		159	litres (L)
	gallons per stroke (gal/stroke)	0.00379	cubic metres per stroke (m^3/stroke)
	ounces (oz)	29.57	millilitres (mL)
	cubic inches ($in.^3$)	16.387	cubic centimetres (cm^3)
	cubic feet (ft^3)	28.3169	litres (L)
		0.0283	cubic metres (m^3)
	quarts (qt)	0.9464	litres (L)
	gallons (gal)	3.7854	litres (L)
	gallons (gal)	0.00379	cubic metres (m^3)
	pounds per barrel (lb/bbl)	2.895	kilograms per cubic metre (kg/m^3)
	barrels per ton (bbl/tn)	0.175	cubic metres per tonne (m^3/t)
Pump output and flow rate	gallons per minute (gpm)	0.00379	cubic metres per minute (m^3/min)
	gallons per hour (gph)	0.00379	cubic metres per hour (m^3/h)
	barrels per stroke (bbl/stroke)	0.159	cubic metres per stroke (m^3/stroke)
	barrels per minute (bbl/min)	0.159	cubic metres per minute (m^3/min)
Pressure	pounds per square inch (psi)	6.895	kilopascals (kPa)
		0.006895	megapascals (MPa)
Temperature	degrees Fahrenheit (°F)	$\frac{°F - 32}{1.8}$	degrees Celsius (°C)
Thermal gradient	1°F per 60 feet	—	1°C per 33 metres
Mass (weight)	ounces (oz)	28.35	grams (g)
	pounds (lb)	453.59	grams (g)
		0.4536	kilograms (kg)
	tons (tn)	0.9072	tonnes (t)
	pounds per foot (lb/ft)	1.488	kilograms per metre (kg/m)
Mud weight	pounds per gallon (ppg)	119.82	kilograms per cubic metre (kg/m^3)
	pounds per cubic foot (lb/ft^3)	16.0	kilograms per cubic metre (kg/m^3)
Pressure gradient	pounds per square inch per foot (psi/ft)	22.621	kilopascals per metre (kPa/m)
Funnel viscosity	seconds per quart (s/qt)	1.057	seconds per litre (s/L)
Yield point	pounds per 100 square feet (lb/100 ft^2)	0.48	pascals (Pa)
Gel strength	pounds per 100 square feet (lb/100 ft^2)	0.48	pascals (Pa)
Filter cake thickness	32nds of an inch	0.8	millimetres (mm)
Power	horsepower (hp)	0.7	kilowatts (kW)
Area	square inches ($in.^2$)	6.45	square centimetres (cm^2)
	square feet (ft^2)	0.0929	square metres (m^2)
	square yards (yd^2)	0.8361	square metres (m^2)
	square miles (mi^2)	2.59	square kilometres (km^2)
	acre (ac)	0.40	hectare (ha)
Drilling line wear	ton-miles (tn•mi)	14.317	megajoules (MJ)
		1.459	tonne-kilometres (t•km)
Torque	foot-pounds (ft•lb)	1.3558	newton metres (N•m)

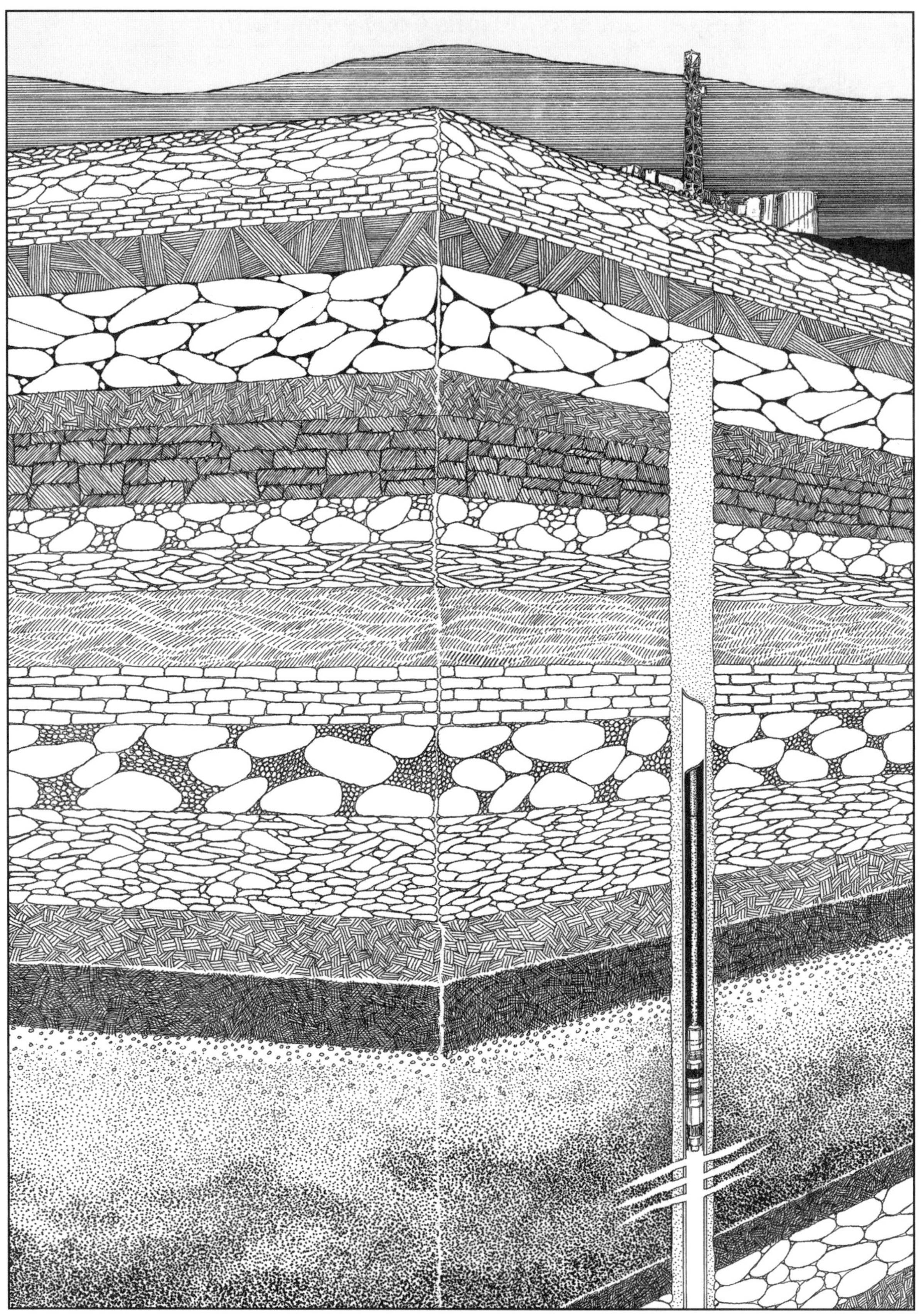

Introduction

Well completion methods vary a great deal because not only are no two wells exactly alike, but also no two producing companies handle a well the same way. It is impossible to talk about the typical well completion because it does not exist. Personal preferences and experience dictate the choices made in selecting individual steps taken during completion. A frequently heard comment is "there are 35 or 40 ways to complete a particular well and five of them are bad, so just avoid the bad ones." Yet, the general completion process—the gathering, checking, and analyzing of data and the choosing, designing, and running of the completion—is the same for every well. Completion planning begins before drilling and continues through completion, stimulation, well servicing, and recompletion. The plan must take much information into account to reach the overall goal of every completion: that is, to produce the most oil and gas at the least possible cost. A well-planned completion accomplishes this goal in two ways. First, it recovers the maximum amount of petroleum (economically speaking) by producing reservoir fluids efficiently. Second, it saves possible later costs of equipment, workover, and recompletion through foresight in planning. No matter who plans a completion, reservoir conditions primarily, along with efficiency and foresight, decide the well's success or failure.

Reservoir Characteristics

A reservoir's characteristics are made up of factors that affect how it will produce its fluids. Reservoir rock types, porosity, absolute permeability, relative permeability to reservoir fluids, saturation, pressure and temperature, and corrosive fluids are included in reservoir data.

Reservoir Rock Types

Reservoirs are almost always made of sedimentary materials. Igneous and metamorphic rocks rarely hold hydrocarbons. Oil-producing sedimentary rocks are generally of two types, clastic and carbonate (fig. 1). It is worth noting that in recent years, operators have made an enormous effort to increase the available natural gas production in the United States. This effort has led to the development of coal bed methane so that today forecasters anticipate coal bed methane will provide upwards of 10 percent of natural gas production in the not too distant future.

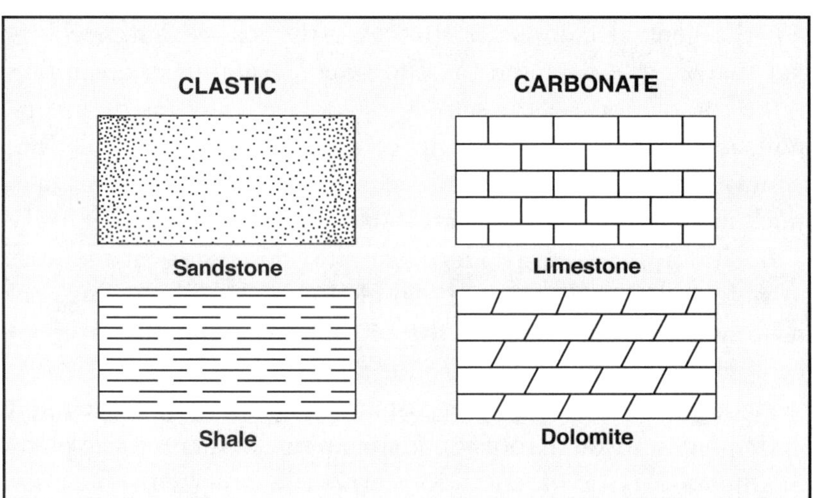

Figure 1. Sedimentary rock types, with standard symbols

Clastic

Clastic rock forms from fragments of rock or shell that are deposited in layers in river deltas, marine channels, and streams along with the organic debris that will become, after millions of years, valuable hydrocarbons. Some scientists believe that hydrocarbons also originate deep in the earth and are trapped in sedimentary rocks as they migrate upwards. The organic theory of the origin of oil and gas states that petroleum formed from the remains of plants and animals. The inorganic theory postulates that hydrocarbons formed deep in the earth and migrated upward. Both theories of the origins of hydrocarbons are possible; consequently, they both are likely to exist. Sandstone is usually a high-quality clastic reservoir rock. On the other hand, shale, because it is usually impermeable, is rarely productive except in areas where commercial amounts of hydrocarbons are obtained from naturally occurring fractures in the shale.

Carbonate

Carbonate rock forms from minerals, mainly calcium carbonate, $CaCO_3$, and calcium magnesium carbonate, $CaMg(CO_3)_2$, that precipitate from seawater over long periods of time. Calcium carbonate (limestone) and magnesium carbonate (dolomite) often come from microscopic sea life, whose fossils are found deposited throughout carbonate formations. Dolomite is produced when at least half of the calcium carbonate in limestone is replaced by magnesium carbonate in a natural chemical process called dolomitization. Another source of limestone and dolomite is from rivers and streams that leach and dissolve inorganic compounds during their courses across the land. In the ocean with the proper temperature and pressure, chemicals fall out of the seawater and are deposited as rocks. Some carbonates also exist in the form of coral reefs where the reefs have been buried under shales or other impermeable rocks. Reefs provide an excellent potential storage place for hydrocarbons.

Coal Bed Methane

The presence of natural gas—consisting mostly of methane—in coal beds has been known about since the advent of subsurface coal mining. Methane was a constant danger to miners both from suffocation and fires; hence, the use of the canary in the cage to determine its presence. Coal beds consist mostly of carbon with some impurities. Methane exists in four environments in coal beds: 1) in cleats, or fractures, 2) dissolved in the water within the cleats, 3) adsorbed onto the coal itself, and 4) absorbed within the coal matrix. Commercial productivity of coal bed methane is governed by the following: depth of wells, pressure, and temperature; degree of fracturing (number and connectivity of cleats); coal type and purity; and thickness of the beds.

Porosity

Porosity is the volume of pore space divided by the bulk volume of solid rock in a formation. It indicates how much fluid a rock holds—whether oil, gas, or water. Reservoir porosity ranges between 3 percent in a tightly bound sandstone to 40 percent in loose, unconsolidated sands. The probable average porosity for any one reservoir ranges between 10 and 20 percent. The greater the porosity, or storage space, the more fluids—oil, gas, water—a reservoir can store. In some coral reef reservoirs, extremely large pores, called vugs, sometimes exist that display porosities in excess of 50 percent.

Several things affect *primary porosity*, the original porous property of a formation. Sands in which grains fit closely are less porous than those sands in which they fit loosely. Porosity is lower when the grains are all different sizes than when they are roughly the same size. Sands made up of jagged grains are less porous than sands made up of round grains. Clay particles and mineral cements, such as calcite (a form of limestone), take up room and clog rock pores, lowering porosity and permeability. Pressure from overlying formations causes reservoir rock to be compacted and reduces space (porosity).

Secondary porosity comes from changes in the reservoir rock, such as those caused by geological pressures, leaching, and dolomitization. Dolomitization occurs when calcium is replaced by magnesium. Magnesium molecules are smaller than calcium; thus porosity is created or increased. Pressure may fracture rock and cause formation fissures where fluids can collect. Water may leach through a formation, dissolving and carrying away minerals. Secondary porosity increases the total porosity of a formation, and it may make the difference between a poor well and a productive one. Porosity, however, is only a part of what makes a well worthwhile. Permeability also plays a big role in reservoir productivity.

Permeability

Permeability, the ease with which fluids move through rock, means more than porosity does to the producing rate of a well. A porous formation may hold hydrocarbons, but if they cannot move to the wellbore, then they cannot be brought to the surface. Porosity, on the other hand, provides space for hydrocarbons to be trapped.

The unit of measurement for permeability is the darcy. A Frenchman named Henri d'Arcy, a nineteenth century hydraulic engineer, studied the permeability of artesian wells. The unit of permeability was later named for him. Petroleum reservoir permeability is usually so small that it is measured in thousandths of a darcy, or millidarcys (md).

Permeability and porosity vary not only from well to well but also within the same reservoir in the same well. Table 1 lists porosity and permeability for several formations in various parts of the U.S. Permeability may be almost nonexistent, less than 0.01 md in tight, fine-grained sandstones and marine limestones; or, it may range upwards of several darcys in loose, coarse, well-rounded sands and rocks. Many formations have permeabilities between 0.1 and 100 md. However, as the quest for production in the world continues, oil and gas finders now obtain commercial production from wells with permeabilities of less than 0.05 md. This improvement is mostly a result of changes in stimulation techniques of these tight, formerly noncommercial reservoirs.

Higher porosity sometimes results in higher permeability; however, pore size and shape and pore connections affect permeability more than porosity. Large pores generally let more fluid pass than do small pores, although a small, round pore may be more open than a large, irregular one. Even large, round pores let very little fluid flow through if they have limited connectivity or if the pore connections are clogged with clay particles or mineral deposits. Large pores with many clean, round connections result in the highest permeability and, as a result, the highest initial production. As a consequence, larger porosity results in higher recoverable hydrocarbons, all other factors being equal. The connectivity of the pores with each other to provide a path for fluids to flow is *tortuosity*. The shorter the flow path compared to a direct line, then the lower the tortuosity. Therefore, a combination of high effective porosity along with low tortuosity results in high permeability, excellent productivity, and high ultimate recovery per unit volume of reservoir rock. The usual unit rock volume used in the U.S. is acre-feet (acres drained times net feet of productive rock). In SI units, rock volume is expressed in cubic metres.

The permeability of a reservoir is an average measurement. Some parts of the reservoir may have bad pore connections—clay, shale, or quartz intrusions—or high tortuosity that lower local permeability. On the other hand, vugs and fissures raise local permeability. All these rock properties affect the total average permeability for the reservoir.

Table 1
Variations in Porosity and Permeability

Formation	Porosity (percent)	Permeability (millidarcies)
Clinch Lee County, VA	9.6	0.9
Dean Sand Dawson County, TX	11.1	0.46
Clearfork Ector County, TX	11.8	<0.01
Second Wilcox Oklahoma County, OK	12.0	100.0
Mt. Simon Sand Cass County, IN	13.8	79.8
Smackover Quitman County, MS	14.1	8.6
Cut Bank Glacier County, MT	15.4	111.5
Bartlesville Anderson County, KA	17.5	25.0
Olympic Hughes County, OK	20.5	35.0
Strawn Cooke County, TX	22.0	81.5
Woodbine Tyler County, TX	22.1	3,390.0
Nacatoch Jack County, TX	24.9	2.6
Nugget Fremont County, WY	24.9	147.5
Serpentine Bastrop County, TX	25.6	0.06
Frio Aransas County, TX	27.1	1,880.0
Eutaw Choctaw County, AL	30.0	100.0

Thickness

Net thickness of reservoir rock is that portion of the rock where permeability and porosity are sufficient to produce hydrocarbons. A reservoir 20 feet or metres thick can produce at twice the rate and contain twice as many hydrocarbons as rock 10 feet or metres thick, if other reservoir rock and fluid properties are the same.

Saturation

Although porosity and permeability are important to the economic productivity of a reservoir, the amount and kind of fluids in the reservoir also play a big role. *Saturation* is a term that describes how much of which fluids a reservoir holds. In talking about a reservoir, the term saturation describes the percentage of total pore space filled by a fluid, whether oil, gas, or water.

The rock that forms most petroleum reservoirs was deposited in an ancient ocean, so it originally held seawater. The oil and gas trapped in it migrated up through permeable rock from deeper regions of greater pressure and temperature where hydrocarbons form. This oil and gas displaced salt water in the pores, lowering their water saturation but raising their hydrocarbon saturation. Average saturation, in a typical oil reservoir that produces oil and gas without water, may be 30 percent water and 70 percent oil. Water saturation of 20 percent and gas saturation of 80 percent may be typical in a gas reservoir. The water that is present in these percentages may be immovable (it cannot flow from the reservoir) water and may be held in place on the rock surface by capillary forces.

Variations in water, oil, and gas saturations cause the fluids to flow differently. Since water is usually less viscous than oil and moves more freely through rock pores, some reservoirs produce more water than oil. Natural gas, on the other hand, is much less viscous than water or oil. Solution gas is gas dissolved in oil. As the oil is produced, the pressure is reduced, and gas comes out of solution. This evolution of gas helps to move oil. However, as the pressure continues to decline, more gas comes out of solution, until, eventually, wells that were originally oilwells might produce gas with little or no oil. Producing mechanisms are controlled by the quantity and physical characteristics of the dissolved gas and the physical characteristics of the oil itself, along with the relative permeability to oil, gas, and water of the reservoir rock. Relative permeability to oil, gas, or water is proportional to the saturations of each of these fluids.

Pressure and Temperature

Pressure and temperature strongly affect the behavior of reservoir fluids. For example, light hydrocarbons such as butane and propane are usually liquid at typical reservoir pressures. When pressure drops as the well flows, these hydrocarbons vaporize and help to carry oil to the wellbore. Knowing the pressure, temperature, and hydrocarbon content of a reservoir is important to enable good decisions for production methods and is especially important for choosing equipment to meet pressure and temperature extremes.

Corrosive Fluids

Along with water, oil, and gas, reservoir fluids may include compounds that are corrosive to completion equipment. (Engineers often call corrosive fluids corrodents.) To avoid expensive replacement of equipment when it becomes damaged or fails from corrosion, the correct equipment must be used to guard against whatever corrodents are found in formation fluids. Hydrogen sulfide in sour wells, carbon dioxide in sweet wells, and brine in older wells often create well performance problems that must be solved ahead of time if production costs are to be kept low and recovery to be kept high.

To summarize—

Reservoir Characteristics

- A reservoir's characteristics include factors that affect how it will produce fluids: reservoir rock types, porosity, absolute permeability and relative permeability to reservoir fluids; saturation, pressure, and temperature; and corrodents.

Reservoir Rock Types

- Reservoirs are almost always made of sedimentary materials, generally of two types, clastic and carbonate. Igneous and metamorphic rocks rarely hold hydrocarbons.

Clastic Rock Formations

- Clastic rock is formed from fragments of rock or shell that are deposited in layers in river deltas, marine channels, and streams along with organic debris that become valuable hydrocarbons after millions of years.

Carbonate Rock Formations

- Carbonate rock forms from minerals—calcium carbonate, $CaCO_3$, and calcium magnesium carbonate, $CaMg(CO_3)_2$—that precipitate from seawater over long periods of time.

Coal Bed Methane exists in four environments within coal beds:

1) Cleats or fractures.
2) Dissolved in water within cleats.
3) Adsorbed onto coal.
4) Absorbed within the coal matrix.

Porosity—is the volume of pore space divided by bulk volume of solid rock in a formation; porosity indicates how much fluid a rock holds—whether oil, gas, or water.

Secondary porosity comes from changes in the reservoir rock, such as those caused by geological pressures, leaching, and dolomitization. Dolomitization occurs when calcium is replaced by magnesium.

Permeability is the ease with which fluids move through rock. Petroleum reservoir permeability is usually so small that it is measured in thousandths of a darcy, or millidarcys (md).

Thickness

- Net thickness of reservoir rock is that portion of the rock where permeability and porosity are sufficient to produce hydrocarbons.

Saturation

- Saturation describes how much of which fluids a reservoir holds and is the state of being completely filled.

Pressure and Temperature

- Knowing the pressure, temperature, and hydrocarbon content of a reservoir is important to enable effective production methods and for choosing equipment to meet pressure and temperature extremes.

Corrodents

- Along with water, oil, and gas, reservoir fluids may include compounds that are corrosive to completion equipment.

Formation Evaluation

In the early days of drilling, when cable tool rigs dominated, the notion of sampling well bailings to find out what the bit was penetrating not only seemed unimportant, but also a waste of time. If a well was going to produce, it would produce—sampling well bailings would never change that.

Today, it is still true that formation evaluation cannot change a well's production potential, but completion techniques suggested by formation data can. Formation evaluation is performed routinely and in more ways than early oil people could have dreamed. Methods for evaluation of subsurface formations now include the sophisticated use of seismic surveys, records from nearby wells, the driller's log, mud logs, core samples, and a multitude of wireline well logs.

Seismic Surveys

Seismic surveys help in deciding where to drill because they give clues to where pay zones may exist (fig. 2). From a seismic record, the location, depth, and size of potential reservoirs can be predicted. Later, if a formation is found to hold petroleum, its volume can be estimated.

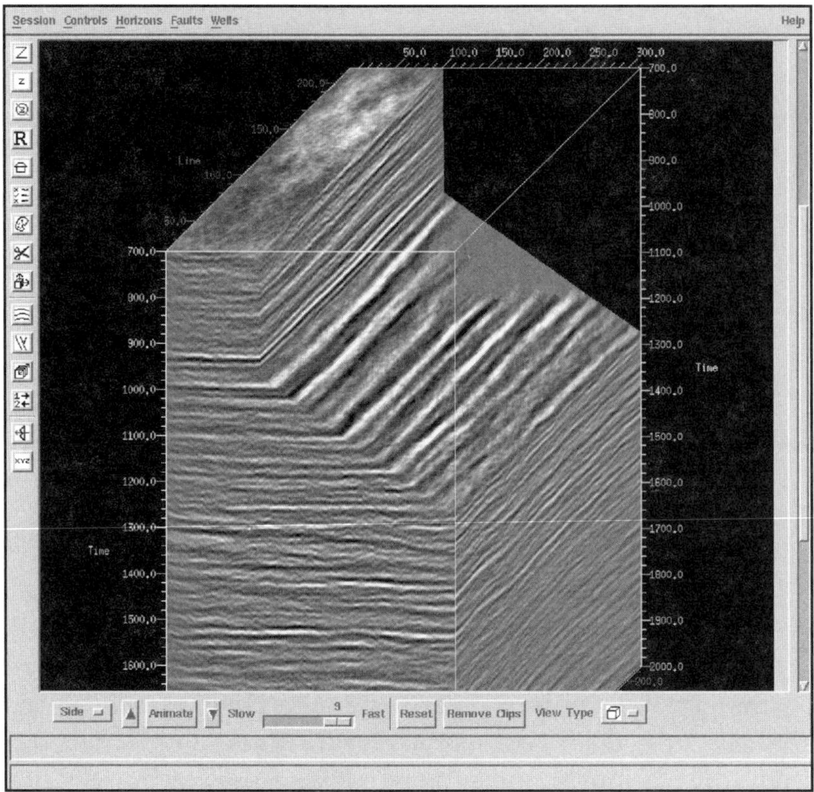

Figure 2. Section from a seismic record (Courtesy of the Bureau of Economic Geology, The University of Texas at Austin)

Records from Nearby Wells

Although formations under neighboring properties may differ greatly, oil and gas fields are sometimes fairly uniform over wide areas. Well records and producing histories from nearby properties can be helpful in evaluating a new well. Clues, such as geological marker beds, may be noted to help locate pay sands; consequently, pressure, temperature, or corrosion problems may be foreseen.

Driller's Logs

The driller's log is an excellent record of formations penetrated by the wellbore. The depth and thickness of each formation bed is noted. Although the driller's log may tell little more about the beds than their relative hardness and type, characteristics can be matched to specific formation types. The driller's log gives the first general picture of the well and provides the basis for later planning.

Mud Logs

Mud logging is a useful evaluation technique that has developed since the advent of rotary drilling. Since mud circulates constantly during drilling, mud logging can provide information about formation constituents more or less continuously.

The mud logger checks mud for oil and gas and collects bit cuttings for analysis. Bit cutting analysis is very useful since it can tell much about rock types and formation characteristics. Information gathered by mud logging is recorded on a continuous graph, called, not coincidentally, a "mud log" (fig. 3). Also, in conjunction with the gathering of samples, a device, sometimes called a sniffer, is attached to the mud-flow stream to detect any increase in the presence of gas in the mud during drilling. Samples are also examined under a microscope using chemicals to determine the presence of oil residue on the cuttings. These methods assist in determining if the wellbore has the potential to become an economic producing well.

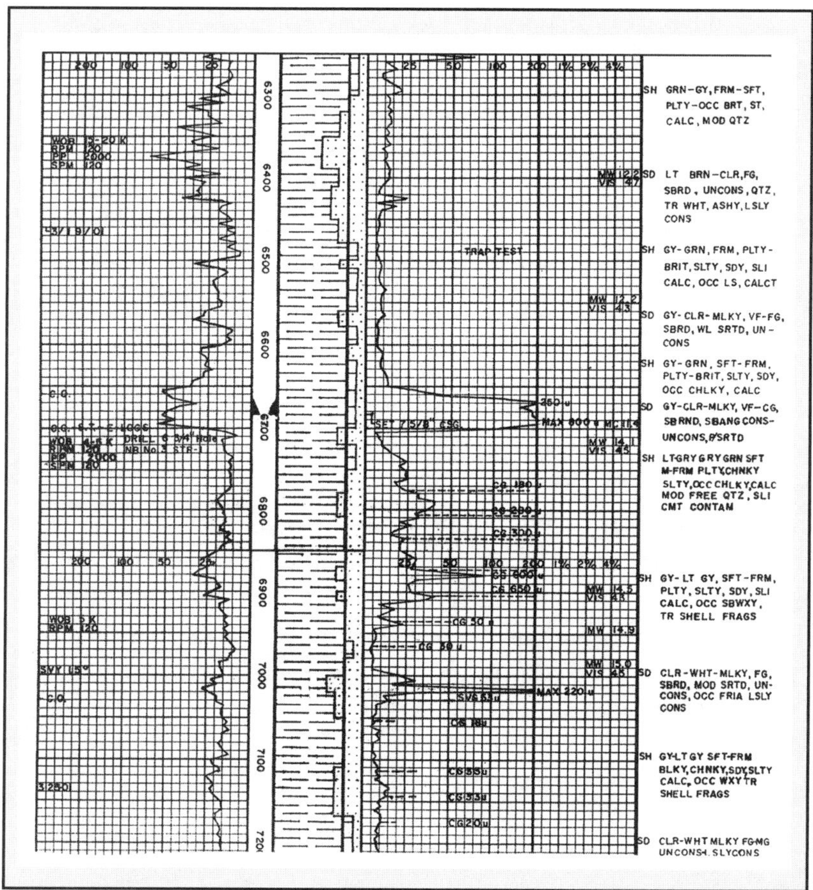

Figure 3. Section from a mud log

Core Samples

If every well could be drilled as one long, continuous core, formation evaluation would be simpler. The rock could be brought up and examined almost exactly as it existed downhole; but coring is very expensive. Instead, a few well planned cores may be taken to tie together and clarify information from the driller's log and the mud log as well as from wireline well logs which are run at a later time. Core analysis is also useful in confirming the rock characteristics, which are measured with open-hole (wireline) logs.

Originally, a short piece of pipe called a biscuit cutter was driven into the bottom of the hole, then jerked out and brought to the surface to obtain a formation sample. Today, most coring is done by barrel and sidewall coring methods. Core analysis has improved steadily over the years.

Barrel or Whole Cores

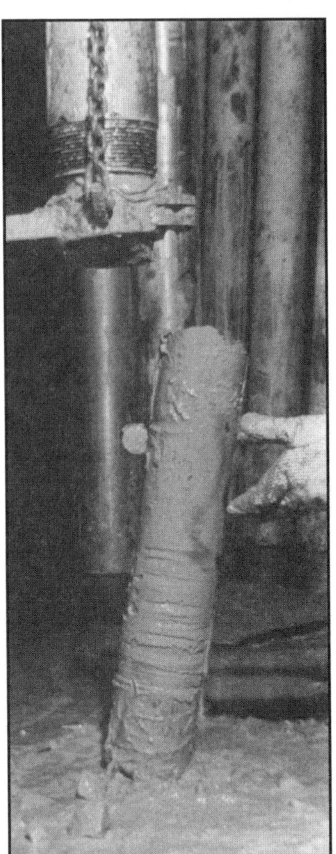

Figure 4. Removal of core

When the bit is about to enter a formation of special interest, a service company may be asked to deliver a barrel core. The coring tool consists of an annular (doughnut-shaped) diamond bit to cut the core and a hollow barrel to catch it (fig. 4). Core samples may be of any length, but cores over 90 feet (30 metres) are difficult to handle. A 30-foot (10-metre) core is average. Core diameters range from 1½ to 5 inches (30 to 130 millimetres), with 5 inches (130 millimetres) being the most popular size for core analysis. Despite its advantages over sidewall coring, barrel coring is done on fewer than 5 percent of all wells cored because it is costly and time-consuming. Also, removing the core from the well is sometimes dangerous since it is possible to swab drilling fluid from the hole and cause a blowout. Consequently, sidewall coring is more commonly used than barrel coring. Whole cores of 5 inches (130 millimetres) in diameter are frequently taken from infill wells drilled for secondary projects where knowledge of reservoir rock properties is especially important and accuracy is needed to assist in design and feasibility of conducting secondary operations. Secondary recovery operations involve injecting fluids, most frequently water, into the reservoir to restore reservoir pressure and to increase the amount of hydrocarbons ultimately produced.

Sidewall Cores

Sidewall cores are taken generally as a follow-up once the well has been drilled and wireline surveys have been completed. A sidewall coring gun is lowered on wireline to the formation chosen for sampling, where it may fire thirty or more hollow bullets into the wellbore wall (fig. 5). The bullets are attached to the gun by cables so that they may be pulled out with the sidewall cores inside them, and retrieved with the gun. The core samples, or plugs, are usually less than 1 inch (25 millimetres) in diameter and up to 4 inches (100 millimetres) long. Since sidewall plugs are small and may also have been altered somewhat by the coring activity, they are not as useful as whole barrel cores for determining porosity, permeability, or fluid saturations. However, sidewall coring is a rapid, inexpensive way to sample a selected formation.

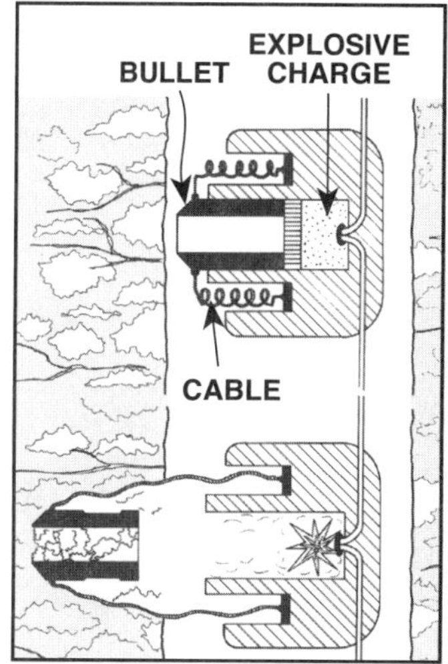

Figure 5. Sidewall coring gun operation

Core Analysis

The well-site geologist and engineer check core samples, whether whole or sidewall, for rock type, physical characteristics, and mineral composition. Fossils help to identify beds. Sidewall cores may be tested for the presence of oil, but they do not reveal as much as whole cores because of their small size and their mud filtrate content. Mud filtrate consists of the water, oil, or both, and dissolved chemicals that flow into the cored rock during the coring process. Mud filtrate also distorts the physical characteristics of the rocks and fluids. Care must be taken in analysis and interpretation of the results of core analysis to determine just how representative the results are compared to conditions that actually exist in the reservoir. The two most important core properties, porosity and permeability, may be estimated using several techniques. Analysis of whole cores may be very detailed, and tests are usually conducted in a laboratory. Core analysis can determine formation properties with a somewhat higher degree of precision compared to other methods. The problem, as with all reservoir evaluation techniques, is that a 5-inch (127-millimetre) diameter core represents only about 3 millionths of an acre (cubic metre). A core is such a very small sample that it could be misleading to wellbore completion planners. Proper well analysis and completion planning require using all available information in decisionmaking.

Wireline Well Logs

Mud logging and core analyses are direct methods of formation evaluation. Wireline well logging (open-hole logging) is the indirect measurement of reservoir rock and fluid properties by electronic, radioactive, and magnetic methods. To log a well, an instrument called a *sonde* is lowered to bottom by wireline (actually conductor line). Electrical, radioactive, and sometimes magnetic properties are recorded, along with hole size, in a logging unit at the surface as the sonde is being lifted (fig. 6).

Open-hole logging companies today gather data in many different ways under many different conditions. Many, but not all, logging devices can be run in a single sonde in one wireline trip. The most useful and economical combination of logs must be chosen to provide the completion people with enough information to determine if a completion should be attempted and to plan a production program for the well. Some data may be taken from logs run on neighboring wells, while other logs are run routinely on every well. Some combinations of caliper logs, spontaneous potential logs, resistivity logs, radioactivity logs, and porosity logs may be included in a typical logging survey. In many instances more than one logging run is required to obtain all of the surveys necessary for operators to make well designed and properly assessed completion decisions.

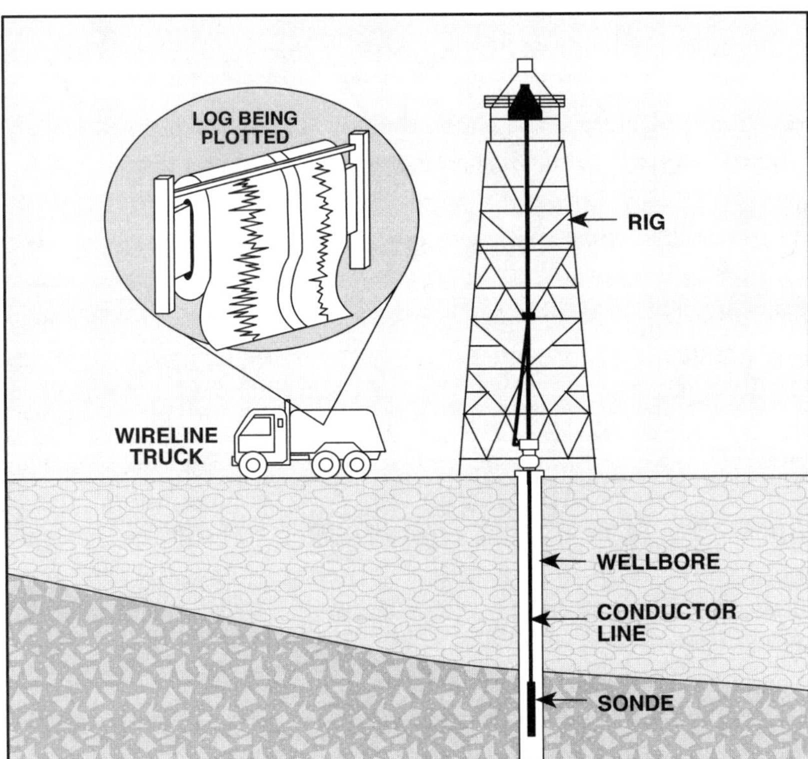

Figure 6. Wireline well logging

Caliper Logs

A caliper is a tool that measures the inner diameter of the wellbore. Pads connected by arms are held against the wall of the wellbore as the device is pulled out of the hole by wireline. Changes in wellbore diameter move the arms in and out and send signals that are recorded at the surface. Wellbore diameter may vary widely due to lateral bit movement, caving formations, mud cake, or flexure (rock bowing into the wellbore due to overburden stress). These variations are important since they affect the interpretation of other log data and provide information for properly designing a cementing program to be used for well completion.

Spontaneous Potential Logs

The spontaneous potential log (SP), along with the gamma ray log, is the most common log. Weak, natural electrical currents that flow in formations adjacent to the wellbore are recorded at the surface. Most minerals are nonconductors when dry, but most are excellent conductors when dissolved in water. When a layer of rock or mud cake separates two areas of differing dissolved chemical concentration, a weak current will flow from the higher to the lower concentration. Usually drilling fluids contain less chemical concentration than formation fluids, which may be very salty. As mud filtrate invades a permeable formation, spontaneous potential causes weak current to flow from the noninvaded to the invaded zone. More importantly, current flows from the noninvaded rock into the wellbore through any impermeable formation, such as shale, above and below the permeable layer (fig. 7).

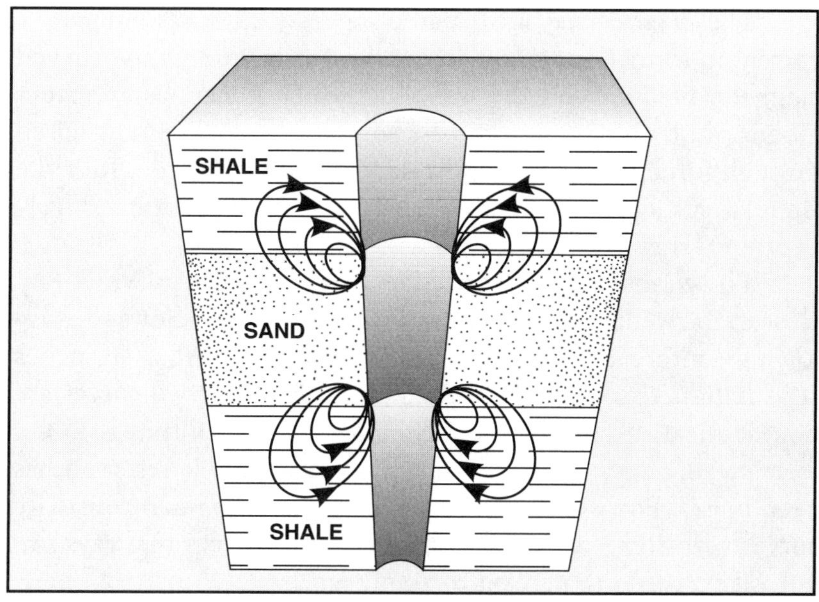

Figure 7. Spontaneous potential around the wellbore

The SP curve is recorded at various well depths in millivolts. The millivolt value is useful in calculations of formation water resistivity. The SP log can be visually interpreted to show formation bed boundaries and thickness, as well as comparative permeability of formation rocks. Because the SP log is so simple to obtain and provides such basic information, it is included in almost every logging run.

Resistivity Logs

Resistivity logs record the resistance of a formation to the flow of electricity through it and is expressed in ohmmeters. Resistance to this flow depends on how much water is present in the pores, the amount and types of chemicals dissolved in the water, the chemical makeup of the reservoir rock, and the amount, if any, of hydrocarbons present in the pores. In this way resistivity is directly related to other formation characteristics. High porosity, high water saturation, and high dissolved chemical content of the water all lower resistivity. Hydrocarbons are poor conductors of electrical current; therefore, oil and gas raise resistivity. If well logs show a formation to be very porous and also to be highly resistive, then the logger may infer that the formation contains hydrocarbons (fig. 8). Common resistivity logs include the lateral focus log, the induction log, and the microresistivity log.

The *lateral focus* log uses a sonde that sends current outward through the rock in a specific pattern. A set of guard electrodes in the sonde focus current sideways into a flat sheet. As the sonde is raised, the sheet of current passes through formation rock. Differences in formation characteristics change the flow of current through the sheet, and these changes are recorded at the surface.

The *induction log*, as its name suggests, involves inducing a current in formation beds. The sonde sets up a doughnut shaped magnetic field around the wellbore, which generates a current monitored by instruments at the surface. As the sonde is raised through formations, changes in the current are logged. Like the lateral focus log, the induction log is very accurate for investigating thin formation beds.

The *microresistivity log* is designed to measure resistivities very close to the wellbore. It consists of two curves: one shows resistivity in the mud cake, the other shows resistivity less than six inches (150 millimetres) into the formation. When the two curves are not identical, an invaded zone is indicated (mud filtrate is in the formation), permeability is probable, and a possible reservoir has been found. Porosity can be calculated from the microresistivity curves with some degree of reliability in instances that effective porosity exceeds 18 percent to 20 percent.

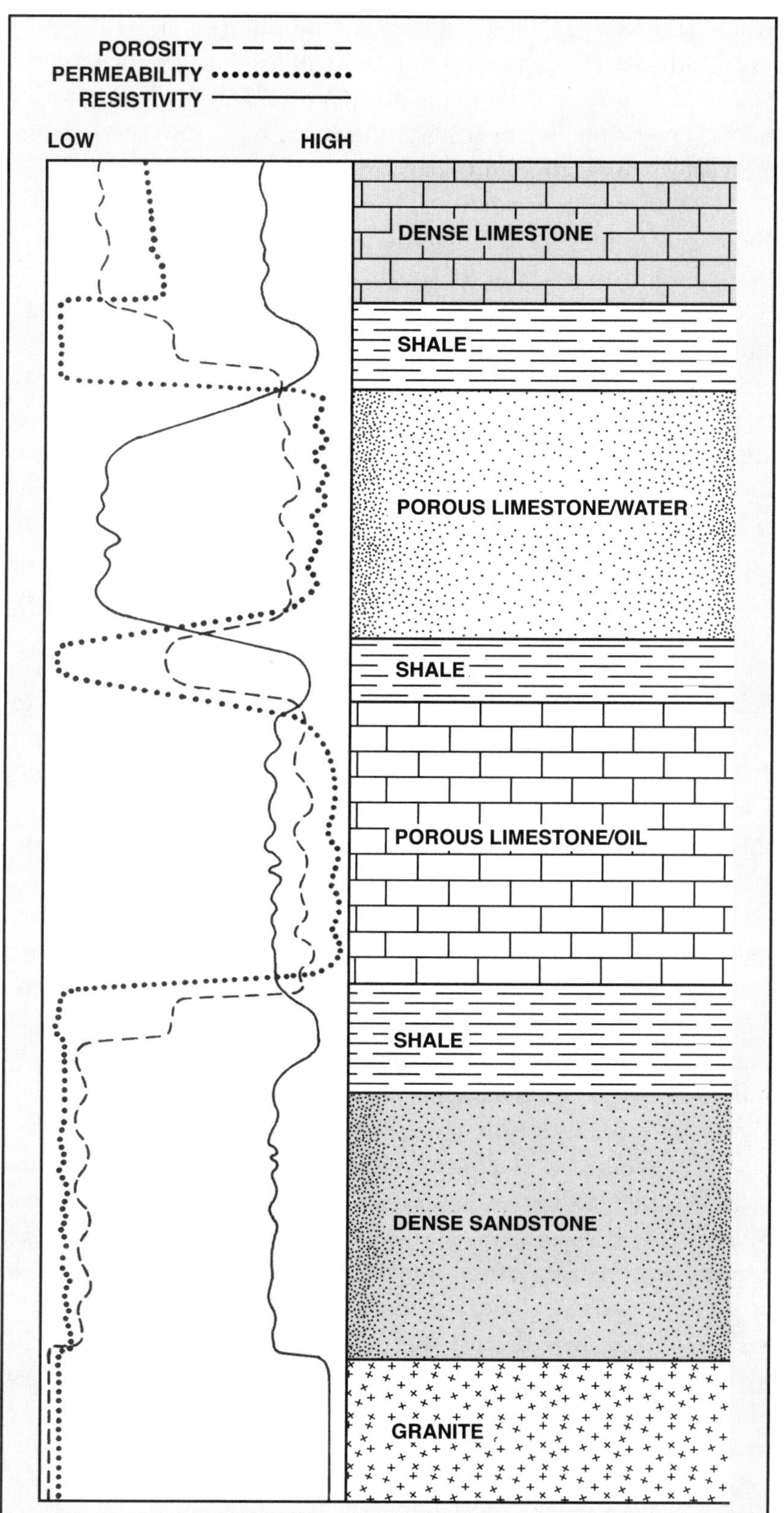

Figure 8. Relationships of porosity, permeability, and resistivity

Radioactivity Logs

Just as resistivity logs record natural and induced electrical currents, radioactivity logs record natural and induced radioactivity. Traces of radioactive elements are deposited in formation sediments. Over time, water leaches them out of porous, permeable rock, such as limestone and clean sandstone, but cannot wash them out of impermeable formations, such as shale and clay-filled sands. Radioactivity logs include gamma ray, neutron, focused density, focused neutron, and photoelectric logs.

Figure 9. Removing a radioactive source from its carrier to load into a logging sonde

Gamma ray logs record natural radiation from the formations. The gamma ray sonde contains a gamma ray detector, such as a Geiger counter. As the sonde is raised in the hole, a curve is graphed in API standard gamma ray units to show gamma ray emissions from the formation beds. Analysis of the curve allows easy location of shales and other rock types and predicts permeability in low permeability formations. The gamma ray log is useful for correlation with the SP log and the neutron log. Just as the SP log is simple to obtain and provides basic information, the gamma ray log is also included in almost every logging run.

The *neutron log* records induced radiation in formation rock. A radioactive source is loaded into the sonde and sent downhole (fig. 9). As neutron radiation bombards the rock around the wellbore, the rock gives off gamma rays from the neutrons it has absorbed. Focused neutron logs used in combination with density logs provide better information to estimate porosity than micro devices do in almost all porosity ranges. Some sondes measure the levels of both gamma rays and unabsorbed neutrons. Other neutron logs can be calibrated to the gamma rays emitted by certain elements such as hydrogen, carbon, oxygen, chlorine, silicon, or calcium. The detected amounts of these elements give information about water and hydrocarbon saturations, salt content, mineral content, and rock types. Neutron logs, used in conjunction with density logs, usually yield reliable porosity readings.

The *density log*, like the neutron log, uses radiation bombardment. The density log responds to bulk densities in formation beds. Bulk density is the total density of a rock, consisting of rock matrix density, fluid density, and pore space volume. Since the densities of sand, limestone, shale, other rocks and minerals, and oil and gas are well known, density log measurements can be converted to bulk density. Mathematical formulas for such rock properties as porosity, hydrocarbon density, and oil-shale yield can be estimated from data taken from density curves.

The photoelectric curve on a log measures the *photoelectric effect (Pe)* on a formation. The photoelectric effect allows loggers to measure known radioactive characteristics of elements and minerals. This curve was fairly recently added to the density log display. From this measurement the type of formation can be identified—sand, shale, limestone, dolomite, or other. For mixtures of rocks and minerals, however, a percentage cannot be directly inferred. The Pe curve helps operators interpret radioactive logs.

Acoustic Logs

Sound travels through dense rock more quickly than through lighter, more porous rock. The acoustic log, also called the sonic log, shows differences in travel times of sound pulses sent through formation beds. Shale and clay, as well as porous rock, slow down the pulses. Using information about formation types from other logs, porosity can be calculated from acoustic logs. Micro (very small) versions of acoustic logs, run in conjunction with special resistivity tools, can assist in determining existence of natural fractures and angle and direction of bed deposition.

Magnetic Resonance Logs

Magnetic characteristics of reservoir rock and reservoir fluids can be measured using a sonde, and from mathematical formulas, porosities and fluid saturations can be estimated. The magnetic resonance tool is similar to density, neutron, and microresistivity tools in that a decentralizer forces the magnetic reading device against the side of the wellbore to provide contact (fig. 10).

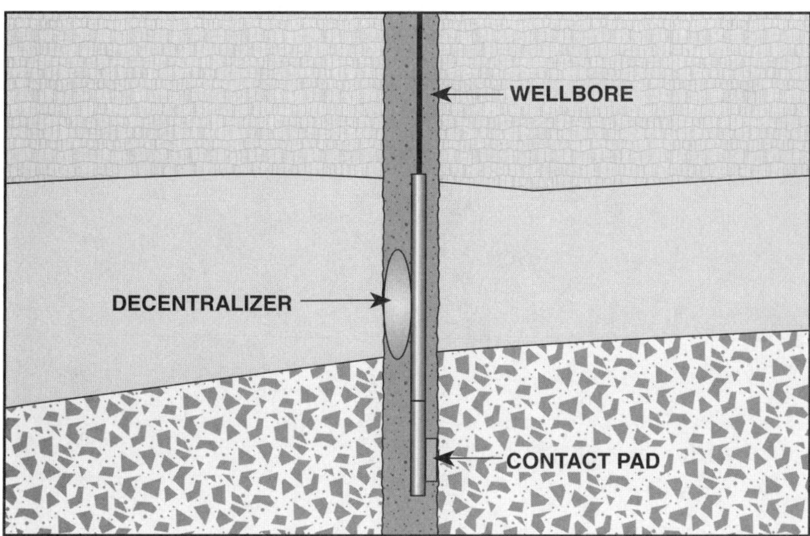

Figure 10. Schematic drawing of magnetic resonance tool

Typical Logging Runs

A wide variety of logs can be taken using a single sonde, but a specific combination is usually chosen for the types of formation data needed. Interpretation of the curves enables the completion people to obtain a picture of lithology, porosity, permeability, and saturation up and down the wellbore. The sonde in (fig. 11) was used to run caliper, SP, induction, gamma ray, neutron, and density logs on a well in East Texas.

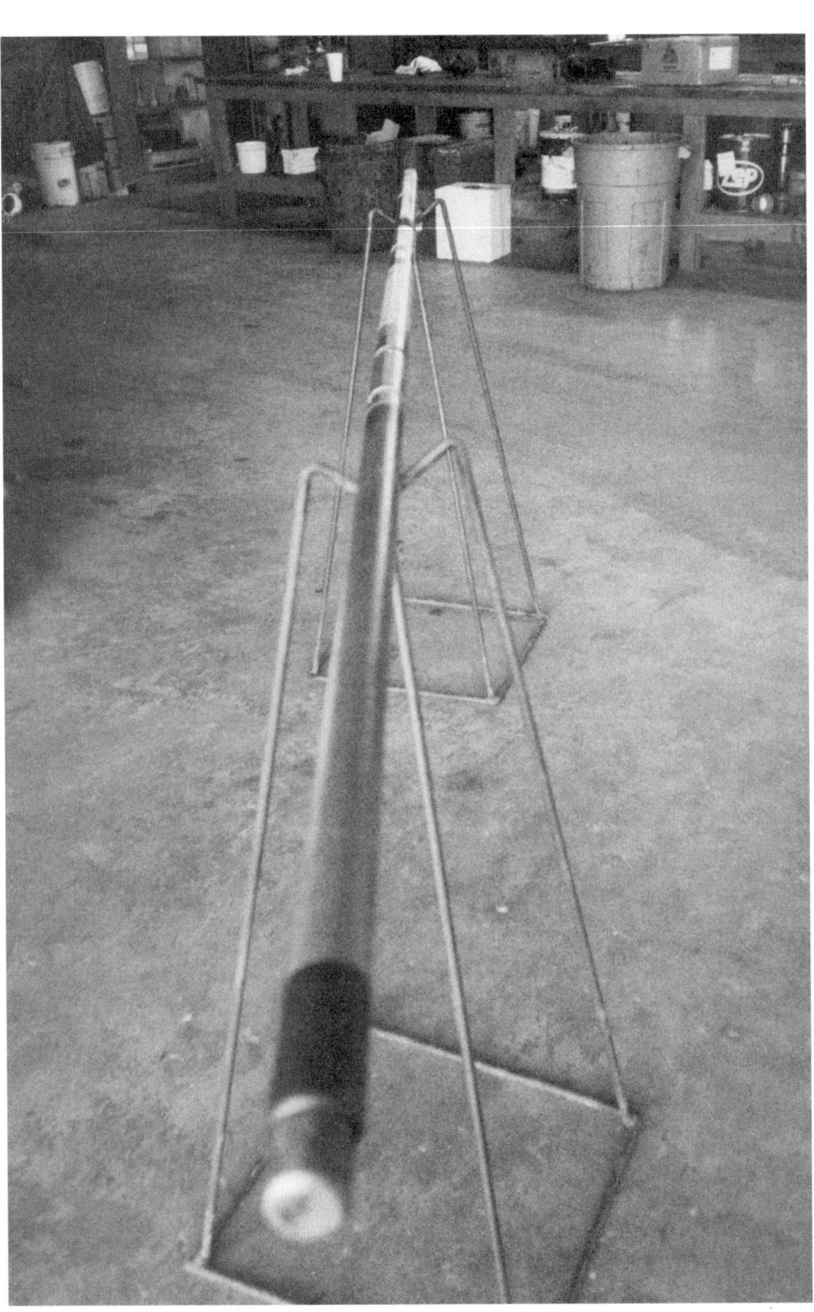

Figure 11. Well logging sonde

Logging While Drilling

In long-reach horizontal drilling, where wireline logging tools cannot be run, logs must be run using drill pipe. Called logging while drilling, such logs have increased the time available to study, plan, and implement completion procedures by providing information while drilling instead of after drilling has been completed. Information obtained from logging while drilling also assists greatly in being able to steer the bit to the desired target.

Formation Evaluation Data

The various formation evaluation techniques—from studying records of neighboring wells to running well logs—are used to find the most promising pay zone or zones. After the well is cased and cemented, these promising reservoirs are perforated, tested, and stimulated if necessary. Formation evaluation has identified the zones that are likely to produce, their rock types, depth, thickness, and gas-oil and oil-water contacts. Formation testing will reveal in detail what fluids the zones contain and what the initial producing rates will be.

To summarize—

Formation Evaluation

- Methods to evaluate subsurface formations include the use of seismic surveys, records from nearby wells, the driller's log, mud logs, core samples, and a multitude of wireline well logs.

Seismic Surveys

- Seismic surveys help to decide where to drill and can also give clues to pay zones. From a seismic record, the location, depth, and size of potential reservoirs can be predicted.

Records from Nearby Wells

- Oil and gas fields are fairly uniform over wide areas. Because of this, well records and producing histories from nearby properties are helpful in evaluating a new well.

Driller's Logs

- A driller's log is an excellent record of the depth and thickness of formations penetrated by the wellbore. This gives the first general picture of the well.

Mud Logs

- Mud logging is an evaluation technique that provides information recorded on a graph, called a mud log, or by gathering samples with a device called a sniffer that detects gas in mud during drilling.

Core Samples

- Cores are taken to determine rock characteristics by using barrel and sidewall coring methods, as well as from the driller's log, the mud log, and from wireline well logs.

Wireline Well Logs

- Wireline well logging is the indirect measurement of reservoir rock and fluid properties by electronic, radioactive, and magnetic methods using a sonde.

Logging While Drilling

- Logging while drilling was developed to gather property information about the wellbore while drilling is taking place and to assist in steering the bit to the desired target.

Formation Evaluation Data

- Formation evaluation identifies zones that are likely to produce by studying records of neighboring wells to running well logs.

Formation Testing

Formation testing may be done before or after running casing, cementing, and perforating. (The few wells that are completed open-hole, of course, are tested only open-hole.) Cased-hole completions are tested through perforations. Open-hole testing followed up by cased-hole testing is also standard in some areas.

The formation test is the final proof of a well's initial capability to produce oil and gas in paying quantities. Cores and logs tell which formations are likely to produce and where to perforate them, but predictions are not the best data on which to base an expensive completion. However, formation tests in general are most useful in reservoirs with medium to high permeability (greater than 15 or 20 millidarcys). In developed areas where log analysis has proved to be successful in identifying potential completion formations, most operators forego the costs of formation testing.

Wireline formation testers and drill stem test tools (DSTs) may be used during or after the conclusion of drilling. Both types of formation tests may be used on a well when considering it for completion.

Wireline Formation Test

A wireline formation test is a quick, inexpensive way to measure pressures at specific depths. Although originally designed to sample formation fluids, this technique is now used for formation testing. The wireline formation tester is actually run on conductor line; the term *wireline* distinguishes this test from those run on drill pipe or tubing. Added benefits of wireline testing are lower costs and less risk than drill stem testing.

The test tool is made up of a rubber pad with a valve in it, a pressure gauge, and testing chambers and sampling chambers interconnected by valves (fig. 12). The tool may be run with a logging sonde or with a bottomhole pressure gauge.

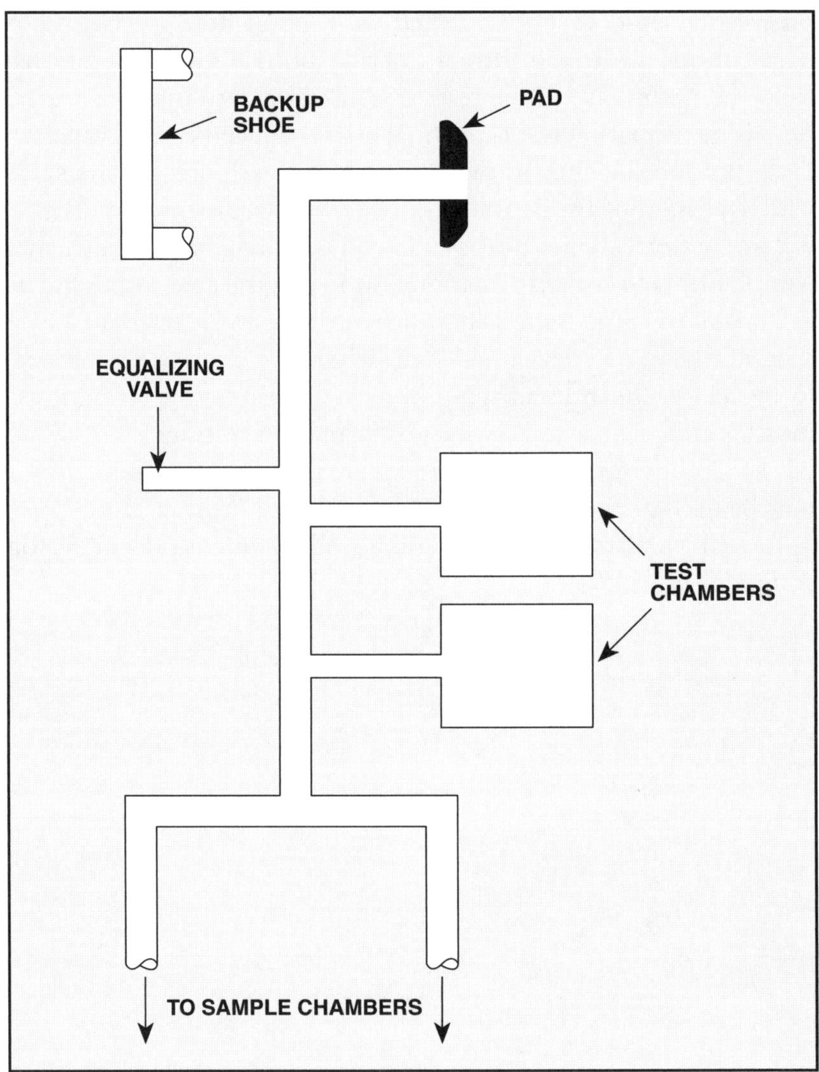

Figure 12. Wireline formation tester design

The zone to be tested is located by wireline depth measurement or by SP log. A backup shoe kicks out to press the pad against the formation sidewall, forming a hydraulic seal from mud in the wellbore. In cased holes, perforations are made into the rock matrix to allow flow into the tool.

The pad valve is opened, and formation fluids enter the tool and register initial shut-in pressure. For a flow period, a test chamber valve is opened and a small piston draws fluids at a steady rate while pressure in the chamber is logged at the surface. A second test chamber is usually opened for a second flow period. The final shut-in pressure is recorded after the second flow period.

Since the test chambers each hold less than an ounce (a few millilitres), fluids drawn into them are almost 100 percent mud filtrate. A sample chamber may be opened to draw a few gallons (litres) of formation fluid. In porous, permeable, well-consolidated formations, a fairly representative reservoir sample may be obtained in the sample chamber.

After a valve is opened to equalize pressure, a getaway shot is fired to release the tool. The tool may then be retrieved, unless it is designed to make more than one test per trip downhole.

Wireline formation tests are useful for investigating oil and gas shows, taking quick readings of hydrostatic pressure and flowing pressure, and estimating permeability. Wireline tests assist in making predictions for zone productivity and may be used in planning more sophisticated formation tests, such as drill stem tests. However, fluid samples are small and tested intervals are thin; therefore, the most useful information usually obtained is formation pressure.

Drill Stem Test

The drill stem test, or DST, like the wireline formation test, was developed as a formation fluid sampling method. It has become a kind of temporary, partial completion of the well that provides data on several feet (metres) or several hundred feet (metres) of formation.

DST Tools

DST tools come in two basic types that may be used for open or cased holes: the single-packer DST tool and straddle-packer DST tool (fig. 13).

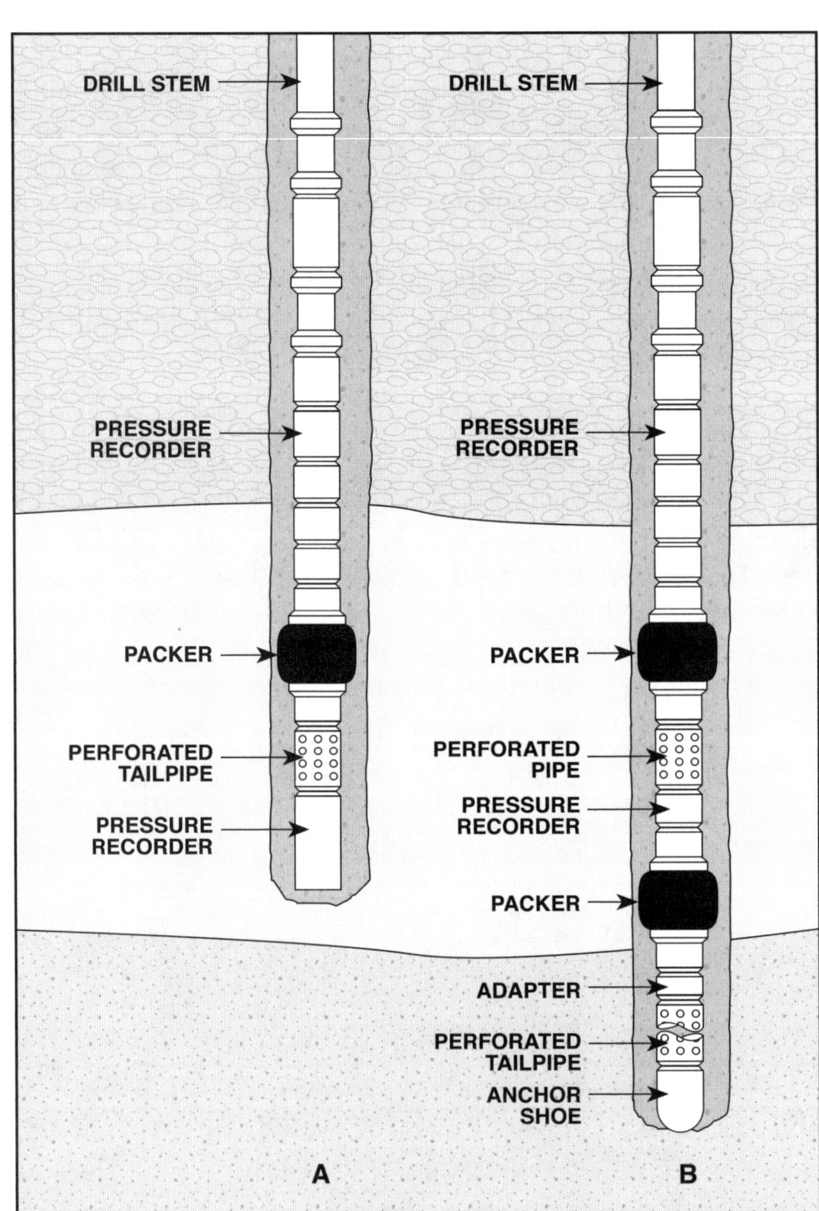

Figure 13. Drill stem test tool assemblies: a) single-packer; b) straddle-packer

The single-packer DST tool isolates formations from below the tool to the bottom of the hole. Perforated pipe is made up below the packer. Formation fluids enter the wellbore and flow through the perforations, through the tool, and up the drill stem to the surface.

The straddle-packer DST tool isolates the formation bed or beds between two packers. The tool is similar to the single-packer tool, in that the lower packer is basically the same as the upper packer except that it is turned upside down. The distance from packer to packer depends on the thickness of the test zone. Formation fluids enter perforated pipe between the packers and flow upwards. In cased holes, where the smooth metal sidewall of the casing may not provide as secure a seat as open holes, DST tools include slips to grip the casing and substitute a sturdier packer to bear the increased set-down weight needed to secure a tight seal.

The inflatable straddle tool consists of two packers spaced exactly as the straddle-packer DST. The only difference is that the inflatable straddle tool is usually set by pump pressure instead of mechanically set.

DST Assemblies

DST assemblies normally contain two pressure recorders. One is usually inside the tool, where it is exposed to pressure inside the perforated pipe. The other is usually below the perforated pipe, where it is exposed to pressure in the annulus. If something goes wrong—if the perforations become plugged, for instance—the two recorders will produce pressure charts with telltale differences. The number and placement of DST pressure recorders are matters to be decided by experience and judgment.

The mechanism used for recording mechanical tests consists of metal plates in the recorders that slowly move under a stylus as it etches a pressure curve. This curve can be analyzed for many kinds of production data.

Digital recorders employ a miniature-recording device that is run into the wellbore along with the DST tool. The sensor consists of a metal diaphragm covered with a quartz coating. Pressure changes cause deflections of the diaphragm and corresponding changes in the frequency transmissions of the quartz crystal. These deflections are converted to pressure readings by mathematical formulas. Pressure performance is read at the surface after the tool has been recovered. Occasionally, the recording device is run on a braided line and pressures are recorded at the surface during the test. Braided line applications are usually confined to offshore and foreign drilling operations.

Temperature readings may also be taken during the test. An instrument records temperature by etching a curve on a metal plate or in the case of digital devices the temperature is recorded in computer bytes. Temperatures, though, vary less than pressures across a field and can be easily calculated from mathematical formulas. Digital devices use an electrical transducer, similar to a transistor or diode, to record temperature. The main goals of drill stem testing are to obtain pressure data and to identify fluids present in the reservoir.

DST Procedures

Once a test zone has been chosen, the mud in the wellbore is circulated and conditioned. Wellbore conditioning gives better DST results and also helps to prevent blowouts caused by the loss of mud weight or by the loosening of a packer in excessively thick mud cake. Cuttings or junk not circulated out of the hole may also damage the DST tool before the test begins.

A cushion of water or compressed gas, usually nitrogen, is often placed in the drill stem when the DST tool is run in. Water is easier to obtain and usually less expensive than gas; consequently, it is more commonly used. If the drill stem were run empty of fluid, the hydrostatic pressure of mud in the annulus could crush the empty drill stem. The cushion supports the drill pipe against mud pressure until the test starts. Cushions can also be drawn off slowly after the tool is opened to prevent formation fluids from flowing so suddenly that the formation rock and downhole equipment are damaged.

After the pressure charts are loaded and the DST tool is laid out and ready to run, the proper length of anchor pipe is tripped in to situate the test assembly at the chosen wellbore depth. All test assembly parts are then added. Sometimes drill collars are made up above the tool as a precaution against collapse if no cushion is used. (Drill collar walls are stronger than drill pipe walls and therefore can withstand higher pressures in the annulus than drill pipe.) Finally, the rest of the drill stem is carefully tripped in.

When the DST tool is in place, the packer seal is set and the test is ready to begin. The tool is opened, and the casing or wellbore annulus is monitored for pressure changes that warn of poor packer seating. Then the cushion, if used, is bled off to reduce pressure so that formation fluids may flow up the drill stem. The strength of this first flow is estimated, either by observing the blow in a bubble bucket (for small wells) or by checking data from a surface computer or other surface readings (for deeper, high-pressure wells). From this estimate the number and lengths of the flow and shut-in periods for the rest of the test are chosen.

Most DSTs include two flow and shut-in periods, the second round being of longer duration than the first. Occasionally tests consist of three rounds. Each round includes a flow period followed by a shut-in period to record the formation pressure. Shut-in periods are usually at least twice as long as flow periods to allow the well to establish equilibrium and to provide information that is useful to the reservoir engineer. The short initial flow assures that the tool is open, clears out any pressure pockets in the wellbore, and removes mud from the drill stem. Produced fluids are caught in a holding tank or burned off as they reach the surface. If they are saved, they may be analyzed later.

When the test is completed, the tool's valves are closed to trap a clean fluid sample and to obtain final shut-in pressures. Then the DST tool is unseated. Fluids in the drill stem are reverse circulated out to keep crude oil or condensate from being spilled over the rig while tripping out. Finally, the drill pipe and tool are carefully tripped out, and the fluid sample and recording devices are retrieved.

DST Interpretation

If the final shut-in pressure varies substantially from the initial shut-in pressure, several overall reservoir characteristics can be inferred. Some of the reservoir characteristics are that the reservoir is depleting, the reservoir is tight or damaged, or that the reservoir is cleaning up. A full analysis of the DST can reveal permeability, reservoir pressure, the real extent of the reservoir, and potential well productivity.

DST data come from two main sources: the pressure records and the fluid sample taken by the tool. Any liquids caught at the surface and saved in a holding tank have probably lost their gas content and are not representative of reservoir gas saturations.

The pressures obtained from the DST are used to calculate pressure behavior plots. The curve is a record of pressure over time (fig. 14). The pressure and time values needed from the DST mechanical charts are taken by using a microscope chart reader, which may be tied in with a computer (fig. 15). Data from digital devices are downloaded to computers for analysis. The plots record pressures that are used in mathematical formulas to describe the well's production potential.

A successful DST shows reservoir pressure; average permeability; the presence and location of permeability changes, such as clay and shale lenses, faults, and pinch-outs; formation damage and production potential without damage; gas-oil and oil-water contacts; pressure depletion rates for small reservoirs; and the radius of investigation, or distance from the wellbore for which the data hold true.

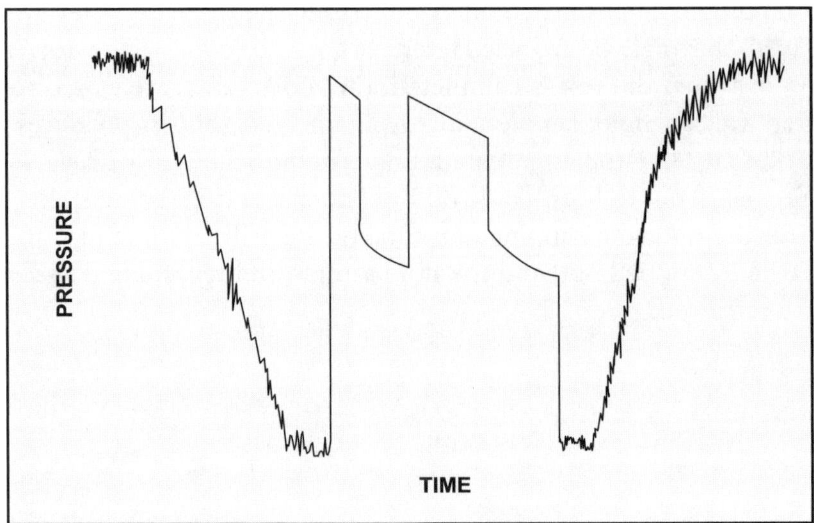

Figure 14. Typical pressure curve from a drill stem test

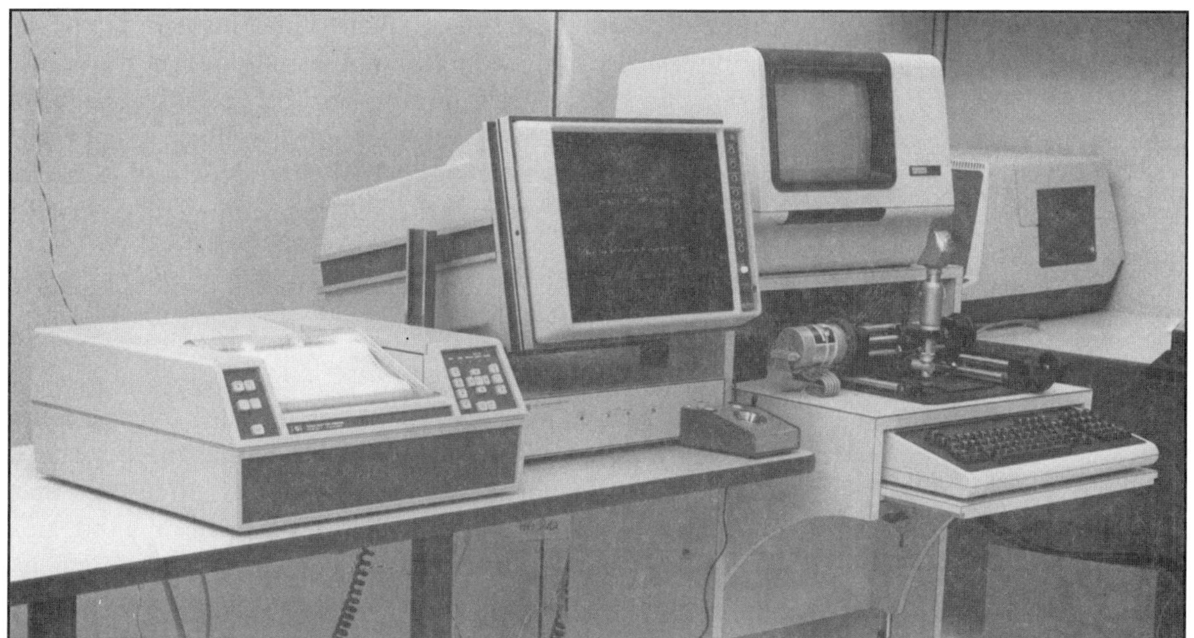

Figure 15. Microscope chart reader with computer

Fluids from the sampling chamber are analyzed for saturation, viscosity, and mineral content, especially salts. A check for corrodents may show hydrogen sulfide or carbon dioxide, which must be considered in the choice of materials for completion equipment. Pressure behavior plots are used in conjunction with the fluid sample to tell how much fluid will be produced, and what kind of fluid can be expected.

Computers are used to graph pressure recordings sent to the surface by conductor line during the DST. The blow and flow periods and shut-in periods can be seen immediately and in great detail, allowing the tester to run a more informative and economical test. Some computers even calculate an immediate pressure behavior plot for the completion engineer.

Since the advent of rotary drilling, DSTs have become a common method of formation testing. The information that they provide is useful in forming a well database for a development field and is valuable for completion planning. Still, more information is sometimes needed to give the most accurate prediction of production rates. Surface well tests can collect more of the needed data.

Surface Well Tests

Surface well tests, the most complete and informative type of formation test, are designed to test not just a zone, but the entire well. DSTs cannot be used in all cases—for instance, on wells with excessively high flow rates, very small wellbores, or caving sidewalls—while surface well tests need only a wellhead to attach to. DST tools may be used with the well-test equipment, especially in open holes where a reservoir may need to be isolated from other zones. Usually, though, a tubing test string is designed and installed after running casing, cementing, and perforating.

Surface Test Equipment

The equipment used to test a well covers a number of basic components and a variety of optional pieces, depending on the testing needs for the particular well. Further, the test equipment that is used is determined by whether the well is open or cased, and whether it is cemented and perforated. The surface test equipment is portable. Portable components can include a tubing head swivel, a surface test tree, surface safety valves, a flow-line manifold, a choke manifold, a heater, a separator, and a tank (fig. 16). Wireline pressure recorders are used to take bottomhole pressure readings if no DST is run with the well test.

A tubing head swivel is a special kind of sub made up on pipe or tubing at the surface. It allows necessary tubing rotation during testing. Tubing attached to it may be rotated without removing the surface test tree. A surface test tree is installed if the well has no permanent Christmas tree. Test components are hooked up to either type of tree for the test. The tree provides flow control for fluids entering the test equipment.

Surface safety valves, either hydraulically or pneumatically controlled, provide added pressure and flow control during the well test.

The flow-line manifold provides points to attach temperature and pressure gauges, as well as outlets for fluid sampling and inlets for chemical injection.

The choke manifold serves as an important pressure control device and a means of reducing the pressure of produced fluids before they reach the separator. It is also where important adjustments are made during the test.

On gas wells, a heater helps to prevent the formation of hydrates, which can clog test lines. If no hydrate problem is expected then the heater is left out of the test package. (A hydrate is an ice-like combination of hydrocarbons and water, which can form in flow lines and other equipment and block flow.)

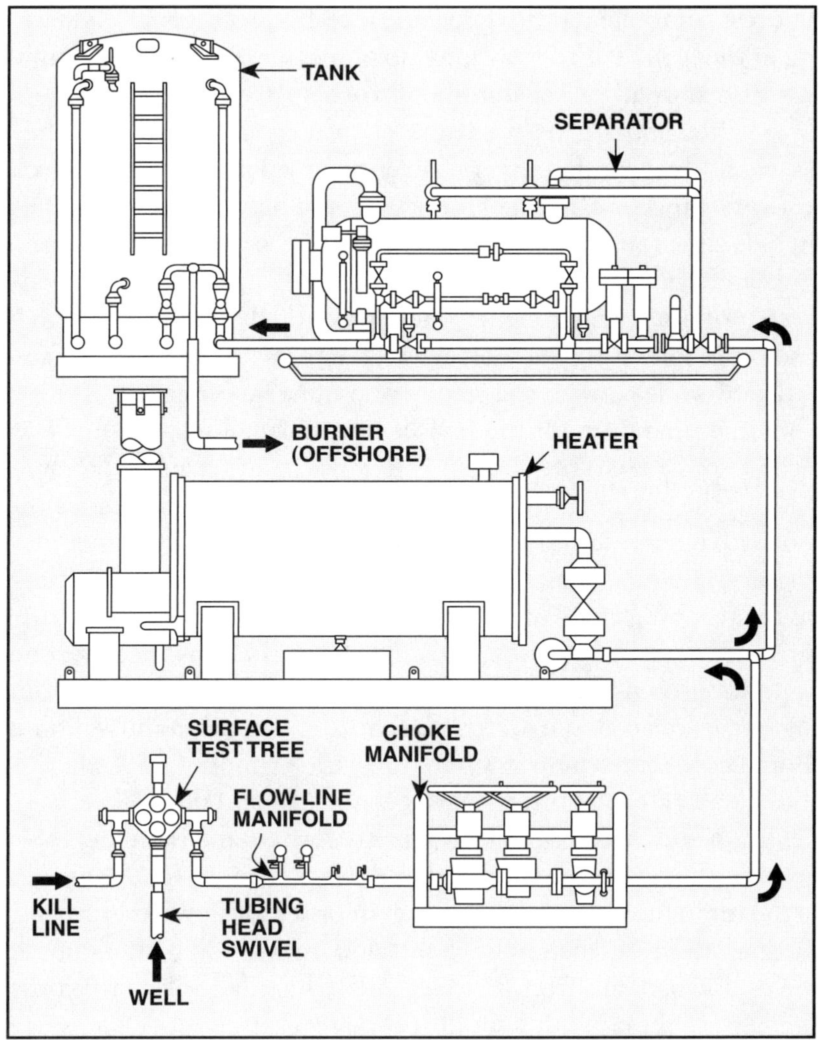

Figure 16. Surface well-test components

The oil and gas separator actually tests the well's production. It includes most of the gauges and recorders that take readings during the test. Where regulations allow, separated water is sent to a pit; otherwise, it is sent to a wastewater disposal system. Oil is sent to a tank, and gas to a flare. In some areas, oil may be sent to a pit to haul away or burn later. In offshore tests, oil and gas are sent to smokeless burners and flares.

Downstream from the separator, a storage tank may be hooked up, generally for two purposes. First, the tank allows the collection of bulk fluid samples for calibration of separator meters against volumes and temperatures measured from the tank. Second, the oil level in a tank can be used to double-check wells whose low flow rates permit the use of tanks.

Surface Test Procedures

The test components are rigged up piece-by-piece, generally starting at the wellhead and working down to the separator. Assembly may also start at the separator and work upstream.

When all parts have been safely placed and connected, they are carefully inspected. Every valve is opened and closed to check it. Every gauge is calibrated for accuracy. These valves and gauges include those on the flow-line manifold, the separator, the storage tank, and auxiliary valves and gauges inserted upstream and downstream of a test component. The burner or flare is also checked. Each component is then pressure-tested to its rated capacity. If any seal shows a leak, then the connection must be redone.

In preparation for the surface test, the well must be made as clean as possible. If production casing has been set and perforated already, the perforating fluid, usually a brine solution, must be removed. To remove it, the well is simply allowed to flow: sometimes for several hours, sometimes for as much as a full day, depending on the test program. Until flow stabilizes and the well is clean, the separator may be bypassed. Flowing the well cleans the test zone of perforation debris and washes mud filtrate out of the reservoir. When the well is closed in after cleanup, the formation and wellbore should contain nothing but pure formation fluids.

When all components have been tested and the well is clean, the test begins. Bottomhole pressure recorders and sampling tools may now be run on wireline. Other downhole tools, such as chokes in the tubing test string, may be run in or set for flow experiments. To start the test, the choke manifold is closed and the required valves downstream from it are opened. Then all valves from the wellhead to the choke manifold are opened. Initial shut-in pressure is read and recorded.

The choke manifold is opened according to a specified plan prescribed by the operator based on the testing company's recommendation. A four-point test, for instance, is commonly used for gas wells.

The choke manifold is slowly and precisely opened to the first setting. Flow pressure is taken. While fluids flow through it, the separator is set up for the test. Separator oil and water levels are selected, and proper back-pressure is chosen. After the flow rate has stabilized, all readings are taken from separator gauges. Then the choke manifold is opened to a second increment, and further adjustments and measurements are made.

Each point at which the choke is reset is followed by a specified flow period, and readings are taken each time. A flow period may last for a few minutes or for many days. At the end of the flow period following the last choke setting, the final flow reading is recorded.

The choke manifold is now closed, and a final series of shut-in pressures is taken until the well has stabilized. Wireline pressure recorders are retrieved. Finally, the well is closed in again. Pressure in the test line is bled off and fluids are dumped into the storage tank or pit or are burned. The test components are then broken down and removed.

Surface Test Interpretation

A detailed well-test report is filled out to use in planning the well completion (fig. 17). Readings from the various gauges and meters at various points in the well-test package are used to fill out a well-test report. These readings may include bottomhole pressure and temperature, wellhead pressure and temperature, casing pressure, static and differential gas pressure, gas flow rate, gas temperature, gas composition, oil flow rate, water flow rate, specific gravities of fluids, and sediment and water content. This information may also be entered into a hand-held computer and transmitted using a cell phone to the reservoir and completion engineers.

Figure 17. Page from a typical well-test report

FORM NO 31-10 (2/74)

BAKER PRODUCTION SERVICES

Well Test Report

COMPANY ______ WELL NO ______
FIELD ______ INTERVAL TESTED ______
COUNTRY ______ BAKER TECHNICIAN ______
DATE ______ BASE CONDITIONS 14 73 POUNDS PER SQUARE INCH AT 60° F PAGE ___ OF ___

BHP GAUGE DEPTH ______
ELEVATION OF ZERO POINT ______
PRESSURES CORRECTED TO DATUM DEPTH ______
BHP GAUGE NO ______ RANGE ______
CALIBRATION NO ______ DATE ______

TUBING SIZE 3-½ WEIGHT ______ LENGTH ______
PACKER TYPE ______ SIZE ______ DEPTH ______
CASING SIZE ______ WEIGHT ______ LENGTH ______
LINER SIZE ______ WEIGHT ______ TOP ______ BOTTOM ______

TIME DATA		CHOKE	SUBSURFACE		WELLHEAD DATA			AMBIENT	SEPARATOR		SAMPLES			GAS	OIL	WATER	GAS	
											GRAVITY							
DATE & TIME HOURS	ELAPSED TIME	SIZE 64THS INCH	PRESSURE PSI	TEMP. °F	TUBING PRESS. PSI	TEMP. °F	CASING PRESS. PSI	TEMP. °F	PRESSURE PSI	TEMP. °F	OIL °API	GAS SP GR	BS&W %	FLOW RATE MMSCFD	FLOW RATE BBL/DAY	FLOW RATE BBL/DAY	OIL RATIO CF/BBL	CHLORIDE PPM
MARCH																		
11,1974			4331	245														
12:00	0	32	4110	"	1200	150	0	85	350	142	35.0	.730	-	4.434	-			16,000
12:30	30	"	3018	"	1205	158	0	85	350	150	35.2	"	2.0	4.436	5204.6	2856.2	852	"
13:00	60	"	3001	"	1140	162	0	86	350	158	35.4	"	1.8	4.313	4240.0	3197.7	1017	"
13:30	90	"	2985	"	1210	160	0	87	350	160	35.7	"	1.6	4.337	5268.3	2447.3	823	"
14:00	120	"	2993	"	1233	165	0	86	350	162	35.2	"	2.1	4.361	5283.8	2638.1	825	"
14:30	150	"	2991	"	1215	162	0	85	350	162	35.6	"	2.0	4.392	5458.5	3053.8	805	"
15:00	180	"	2994	"	1218	160	0	85	350	162	35.0	"	2.3	4.392	5230.8	3051.1	840	"
15:30	210	"	2999	"	1210	165	0	85	350	160	35.2	"	2.2	4.337	5489.6	2931.5	790	"
16:00	240	"	3001	"	1214	165	0	85	350	160	35.0	"	1.7	4.306	5475.2	2854.7	786	"
16:30	270	"	3005	"	1219	165	0	85	350	160	35.6	"	2.1	4.337	5410.6	3106.5	802	"
17:00	300	"	3001	"	1212	165	0	85	350	160	35.4	"	1.6	4.369	5863.1	2980.1	745	"
17:30	330	"	3001	"	1215	165	0	85	350	160	35.0	"	1.8	4.369	5300.0	2453.9	824	"
18:00	360	"	3000	"	1218	165	0	85	350	160	35.0	"	2.1	4.369	5706.5	2872.8	766	"

Bottomhole pressure and temperature are taken from wireline recorders run during the test. Wellhead pressure and temperature are taken from gauges on the surface test tree. These readings are used to prepare pressure behavior plots for the well-test report.

Casing pressure is taken from a wellhead gauge. Casing pressure is the pressure in the casing-tubing annulus. Leakage around a packer usually shows up as rising casing pressure readings during testing. Tubing pressure is also monitored. When annular production is a possibility, a packer may not be run on the tubing test string, and casing pressure is carefully recorded.

Static and differential gas pressures are taken from meters on the separator. Static gas pressure is the pressure exerted by the flowing gas. Differential gas pressure is the measure of the drop in static gas pressure across an orifice plate in the separator's gas meter run. These two readings are used to calculate the gas flow rate.

Gas temperature is taken from a gauge on the gas meter run alongside the separator. Gas temperature is important, since the temperature of a gas has a great effect on the calculation of its volume. Hydrate formation, a problem in gas wells and wells with high gas-oil ratios, results from substantial temperature and pressure drops.

Natural gas includes methane and other light hydrocarbons, and may also contain carbon dioxide, hydrogen sulfide, nitrogen, and other contaminants. Gas composition is measured in one of two ways: by mass spectrometer or by gas chromatography. The mass spectrometer gives the specific gravity of a gas, liquid, or solid, which can be used to identify gas samples. Gas chromatography is often linked with a computer to provide a quick and accurate method for analyzing the composition of gases retrieved by the separator in a well test.

The oil flow rate is read from one or more oil meters on the separator. Some separators have one oil meter. Others combine readings from two or more meters to arrive at the total oil flow rate.

The water flow rate is measured after gas and oil are separated. The water meter takes volume readings as the water leaves the separator for disposal.

The specific gravities of oil and gas produced by the well may be measured from bottomhole samples taken at the start of the test. The API gravity of crude oil is an important part of its rating and composition and is used to assist in determining reserves and market value of the crude. In gas wells, the specific gravity and composition of the gas are necessary to accurately estimate gas reserves. In oilwells, the specific gravity of produced gas helps in calculating its volume and solubility. Gas-in-oil solubility must be known for stock tank measurement, reserve determination, and refinery processes. Heating value and composition help to determine the marketability of and price to be received for the gas.

Sediment and water (S&W) include everything that is not salable oil and gas. The separator and heater-treater remove most of it from produced fluids, which is necessary to bring the remaining percentage down to levels set by the S&W purchaser and governmental regulations.

Calculations are made using values from many of the readings taken during the test. The solutions to the equations tell much about the production potential of the well.

Usefulness Of Test Data

Wireline formation tests, DSTs, and surface well tests may be used alone or in combination, depending on how much money and time is to be spent in gathering formation test data and on the kind of data needed.

The tests vary in scope and duration. A wireline formation test measures the pressure of a small area for a few minutes. A DST measures the pressure of a selected zone for a few hours. A well-test package is installed essentially as an experimental fluid processor that tests flow characteristics of the whole well for as short as a few hours or as long as several weeks.

Usually, the longer and more comprehensive the test, the more useful are the formation data. Better data allow better decisions in completion design and production planning throughout the well life. Bringing a property into production may easily cost several million dollars, but even a smaller investment cannot be recovered by faulty completions based on insufficient or incorrect data. Thus longer, more expensive tests are sometimes worthwhile.

Generally, formation testing tells what fluids will be produced and at what rates they will be produced. This general information is put together from many details that may vary widely between any two particular wells, even if they produce similar fluids at similar rates. Formation test data, along with logging, coring, and other data that have been gathered from a specific well, help to determine which completion is most suitable for the well.

As less permeable, highly fractured reservoirs are drilled and completed in the United States, DST and wireline tests are less frequently conducted. Operators rely on the multitude of openhole log suites available to make the decisions to run casing or plug and abandon.

To summarize—

Formation Testing

- The formation test is the final proof of a well's initial capability to produce oil and gas in paying quantities.

Wireline Formation Test

- A wireline formation test is a quick, inexpensive way to measure pressures at specific depths.

Drill Stem Test

- The drill stem test (DST) is a temporary, partial completion of the well that provides data on several feet to hundreds of feet of formation.

DST Tools

- Single-packer DST tools isolate formations from below the tool to the bottom of the hole.
- Straddle-packer DST tools isolate the formation bed or beds between two packers.
- Inflatable straddle tools consist of two packers spaced like the straddle-packer DST and set by pump pressure.

DST assemblies contain two pressure recorders: one inside the tool (where it is exposed to pressure inside the perforated pipe), and one below the perforated pipe (where it is exposed to pressure in the annulus).

DST Procedures

- Wellbore conditioning gives better DST results and helps prevent blowouts.
- DSTs include flow and shut-in periods to record formation pressure.
- DSTs provide useful information to form a well database for a development field and is valuable for completion planning.

Surface Well Tests—are designed to test a well by using surface test equipment, surface test procedures, and surface test interpretation.

Completion Design

No two wells are alike. Every good completion design takes into account the uniqueness of a well. It would be difficult, and hardly helpful, to look at every possible completion design, but three items are part of every design process: 1) the completion method most suitable for the well, 2) the completion system chosen to bring the fluids to the wellhead, and 3) the completion cost versus production potential.

Completion Methods

The completion method chosen greatly affects the communication from the reservoir to the wellbore. Basically there are two methods: open-hole completion and perforated completion. In conventional perforated completions, reservoir fluids communicate to the wellbore through perforations (holes) made in the casing and cement. Perforating must be delayed until the cement has properly set and the bond log has been run. In open-hole completions, casing is set and cemented at or near the top of the production interval and the reservoir is drilled with a bit that is smaller than the inside diameter of the casing. Coal bed methane wells and horizontal wells are often completed in open hole.

Open-hole Completion

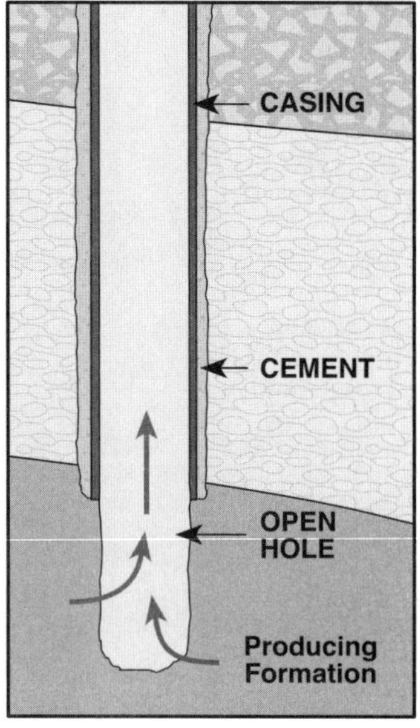

Figure 18. Open-hole completion

The formation face is bare rock, allowing uninhibited flow in large volume up the wellbore diameter. However, an open-hole, or barefoot, completion is used only in special cases: (1) in thick, continuous, well-consolidated carbonate reservoirs (fig. 18), (2) in horizontal wells because of the difficulty of obtaining primary cementing plus the formations may be exposed to open hole for great distances, which increases a well's productivity, and (3) in coal bed completions, wherein the entire coal bed seam is exposed to the wellbore and possibly enlarged by underreaming or explosives.

Ideally, in an open-hole completion, the zone should be thick enough to produce a large fluid volume, have no interspersed water reservoirs that could dilute produced crude, and produce no sand or expose a formation that could cave into the wellbore. Pressure and flow rates may be hard to control in open holes. Artificial lift and well stimulation may pose problems later in the well's life that outweigh the initial advantage of low completion costs and high initial productivity. Although open-hole completions are common in some areas, the total number of new wells completed with open-hole methods, except for coal beds and horizontal holes, is small.

Perforated Completion

The majority of wells are cased and cemented through the producing zone and then perforated (fig. 19). Casing prevents rock caving and provides protective housing for completion tools and subsurface artificial-lift equipment. Cement fills the annular space between the rock and the casing, supports the casing and prevents its external corrosion, and stops fluid migration between neighboring formations. A carefully calculated number of perforations made at precise depths provide wellbore communication at controlled pressure and flow rates. The advantages of perforated completions include wellbore support and protection; control of gas influx (in oilwells); control of water influx in all wells; and the control of well stimulation.

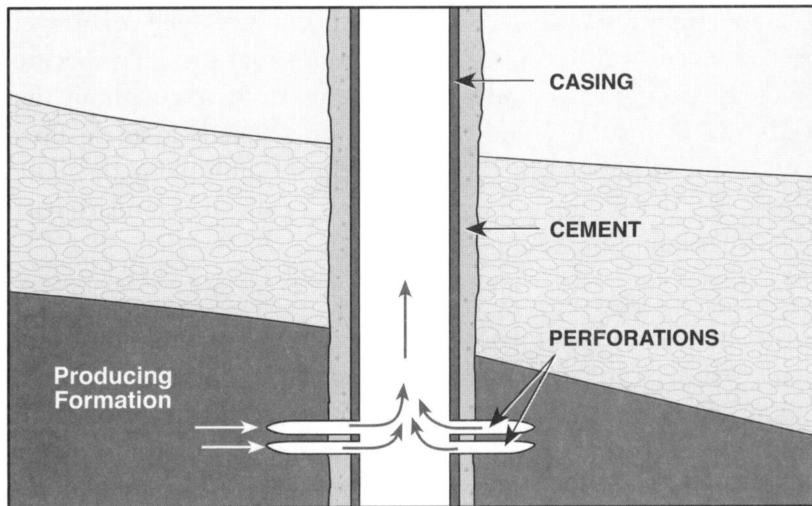

Figure 19. Perforated completion

Screen Liners

Some wells could be completed open hole if the bare producing formation were competent—that is, consolidated and not made up of loose sand or rock fragments. Formation competence is the ability of rock to resist breakdown from the erosive flow of produced fluids. This competence may be analyzed from core samples and also predicted from the amount of sediment in formation test fluid samples. An incompetent formation completed by means of an open hole is likely to fill up with sand or shale, which can restrict or terminate production and require expensive clean out of the wellbore. To avoid the expense of running an entire casing string or cemented liner, a common solution in wells completed in unconsolidated sands is to run a screen liner, often with a gravel pack. The screen liner is a length of perforated pipe, usually wrapped with special wire (fig. 20). The wire is closely spaced to allow passage of fluids but not of sand. To provide a further barrier against the movement of sand into the wellbore, a gravel pack may be placed around the screen liner. Gravel is actually round sand or glass beads of even size and coarse mesh. A 3-inch (75-millimetre) thickness of gravel placed around a wire-wrapped screen liner may keep a well flowing for several years. A gravel-packed screen liner may also be placed in cased wells that have begun to have sand problems. Screen liners are also used in steam floods where producing wells that have been steamed may produce substantial amounts of sand along with viscous crude oil. Screen liners often can be retrieved inexpensively, using sandlines or tubing, then cleaned, repaired, and reinstalled.

Figure 20. Wire-wrapped screen liner

Flow Paths

Whether open hole, cased, perforated, underreamed, or gravel packed, every well has one or more zones that produce specific fluids at specific rates under specific conditions throughout the well's producing life. One producing zone needs a single flow path, while two or more zones may require multiple paths. The advantages of running tubing into the well make it a standard part of most completion designs.

Single Paths

Either production casing or tubing may provide a conduit for well fluids to flow from the formation to the surface. Producing through tubing offers many advantages. Tubing damaged by erosion from sand or by corrosion from carbon dioxide, hydrogen sulfide, or salt can be easily replaced. Damaged casing is more difficult and more expensive to repair. Tubing may also be made of corrosion-resistant materials, and erosion-resistant tubing subs may be made up at key points where sand erosion is expected to be severe.

Tubing may also serve other purposes throughout the life of the well. It acts as a tool for circulation, either for killing the well or for cleaning it out. Well pumping methods also generally need a tubing string, which provides not only a conduit for pumped liquids but also a production casing-tubing annulus that serves as a gas separation chamber. Gas separates from the oil and collects in the casing-tubing annulus to be sold. Since pumping methods are designed for liquids, separating the gas from the oil before the oil reaches the pump increases pumping efficiency.

Mature gas wells that have experienced substantial decreases in reservoir pressure may experience loading. Loading occurs when water and hydrocarbons condense (become liquids) as the flow stream goes from higher temperatures and pressures to lower temperature. Note that the corresponding pressure decrease is usually not sufficient to counteract this condensation. Loading can cause a well to "die" and require frequent shut-in or swabbing to restore to production. A plunger lift device can be used to help overcome the failure to flow. By shutting in the well for a time, the plunger falls near the bottom of the tubing. A tubing stop or a seating nipple prevents the plunger from falling out the end of the tubing.

When the well is reopened, the plunger forms a seal between the bottom of the liquids accumulated in the tubing and the incoming reservoir fluids. This procedure allows the accumulated liquids to be produced to the surface, along with gas and entrained liquids from the reservoir, until the cycle is repeated. A surface controller can be set to allow a few or many cycles per 24 hours as dictated by the well's producing characteristics. Plunger lift installations work much better in instances where the casing-tubing annulus can be used as a storage compartment while the plunger is falling to and resting on the tubing stop.

Multiple Paths

When more than one zone is produced through a single well, a flow path may be designed for each zone, with production rate, production quality, and sizing considerations the same as for single paths. Other than cost, the major added question in multiple completion is whether there is enough room for multiple flow paths. In a dual completion, for instance, both tubing strings may be sized down to fit the wellbore. Or, one zone may be depleted through normal-size tubing while the second zone is shut in or completion is delayed. After the original completion is abandoned, the second zone is opened to production. Some single-tubing designs isolate multiple zones with packers and provide each zone with a circulation device that may be opened or closed to select the producing zone. Still other designs merge several zones into one and allow fluids to commingle as a single flow. Each multiple completion plan meets the challenge of multiple producing zones in different ways and at different costs. The choice of plan depends on wellbore and flow path sizes, the particular design's cost effectiveness, and governmental regulations.

Completion Systems

After the flow path or flow paths have been selected, a completion system is designed for production. A completion system is an assembly of tools designed to meet specific needs for efficient production of the well.

Conventional Systems

Conventional completions are cased, cemented, and perforated completions with tubing and often with one or more packers. The most common conventional system is the single-zone completion with tubing and packer (fig. 21). The packer is set above the perforated zone to seal off the casing-tubing annulus above it and to direct all produced fluids to the wellhead through tubing. Single-zone conventional completions make up the great majority of all completed wells.

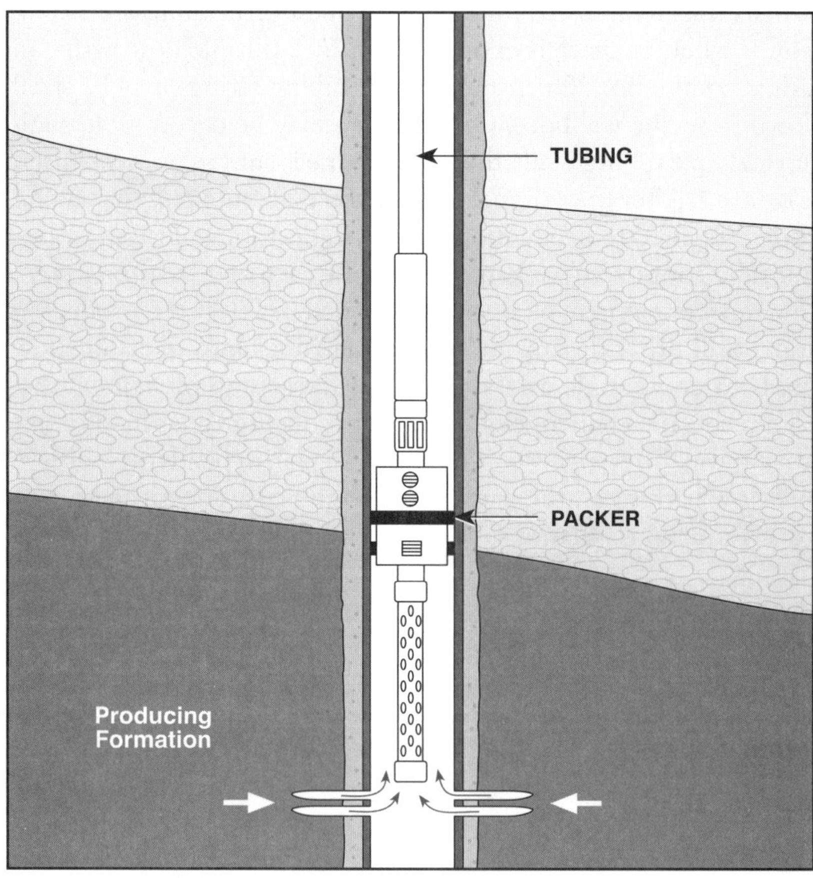

Figure 21. Conventional single-zone completion

Multiple-zone conventional systems are based on the single-zone design. In the case of two producing intervals, one option is to send fluids from one zone up tubing and fluids from the other up the casing-tubing annulus. This system is called a single-string dual completion. The weaker of the two zones (the one with less pressure), which can be expected to need artificial lift sooner, is produced through the tubing to accommodate pumping when required. If the lower zone is the weaker one, then a packer is set between the two zones, and the lower zone produces through tubing while the upper zone produces through the annulus (fig. 22).

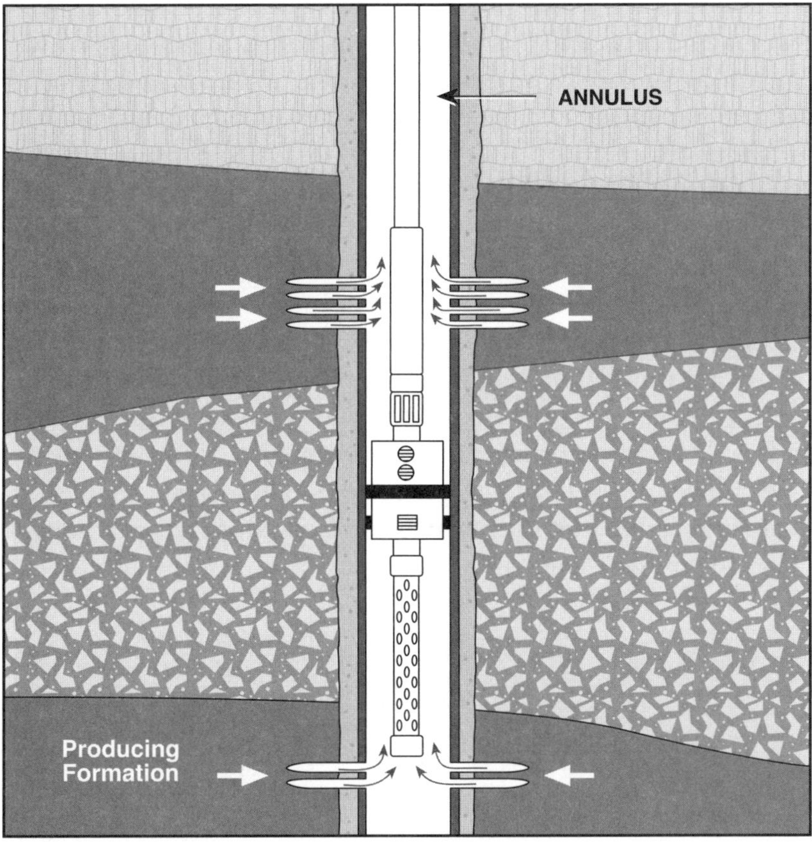

Figure 22. Single-string dual completion: lower zone through tubing; upper zone through annulus

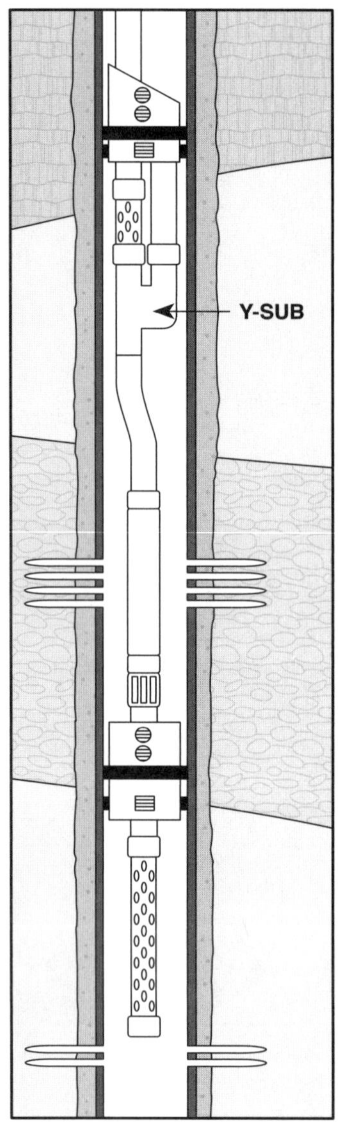

Figure 23. Single-string dual with Y-sub: upper zone through tubing

If the upper zone is the poorer producer, then the flow is directed to the tubing through the use of a *Y-sub* (fig. 23). Fluids from the lower zone move past the upper zone inside tubing, then switch to the annulus; meanwhile fluids from the upper zone move through the annulus and then switch to the tubing. The lower zone can also be produced through the tubing if this kind of production later becomes more efficient for the well.

If one zone primarily produces oil and the other primarily produces gas, then the oil zone is produced through the tubing and the gas is produced though the casing-tubing annulus. This arrangement provides for maximum lift efficiency. The drawback in these systems is that the casing is exposed to produced fluids that may be corrosive. Also this system is difficult and somewhat dangerous to use for reservoirs with extremely high pressures because the casing may be exposed to forces that could cause it to burst.

Another option with multiple reservoirs is to install multiple tubing strings. This system is called a *parallel string* completion (fig. 24). Packers are used to isolate zones, and tubing strings are run from each packer to the wellhead. It is possible to have as many as four parallel tubing strings in a single well, but dual parallel strings are the most common. Disadvantages of multiple tubing strings are that flow paths must be sized down to fit the wellbore and well repairs can become very complicated and expensive. The choice between single string and parallel string systems for dual completion comes down to total costs and economic and mechanical experience of the operator.

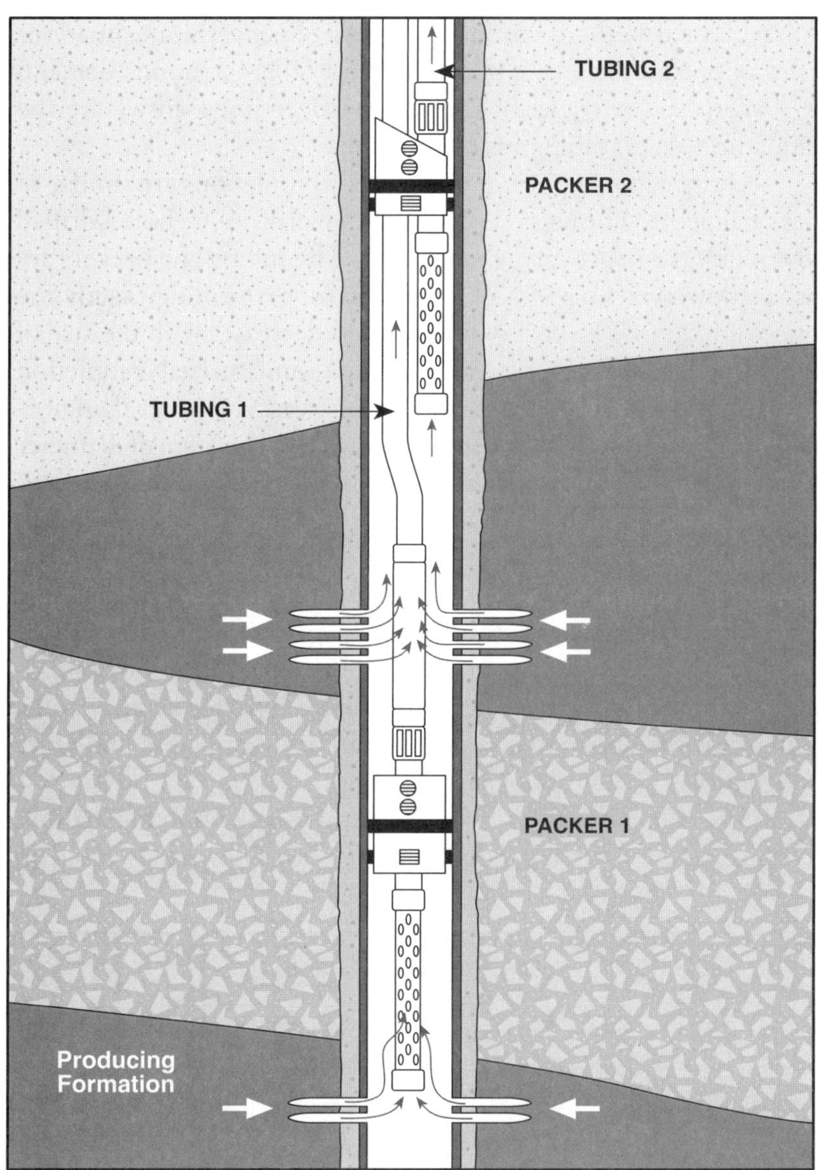

Figure 24. Parallel string completion

Specialized Systems

Conventional methods are not always cost effective. Some situations offer opportunities for cost cutting, and other situations demand it. Two specialized systems that are relatively inexpensive are the miniaturized and tubingless completion.

As the name implies, miniaturized completions are similar to but smaller than normal completions. A small diameter hole is drilled that requires smaller casing than usual. The tubing and packers are scaled-down versions of regular size tools, performing exactly the same functions. The cost of a well often decreases with the size of the rig, wellbore, and completion tools. A miniaturized completion may be used whenever it is possible to use casing smaller than normal in diameter, such as in a low volume well where small diameter tubing can efficiently produce the well. Normally, a miniaturized completion occurs when production casing is smaller than 4½ inches (114.3 millimetres) in diameter. Miniaturized systems have been used in multiple completions where several small diameter casing strings are cemented into a single hole (fig. 25). The single wellbore penetrates all reservoirs to be produced, but separate production casing, tubing, and packer are provided for each perforated pay zone. A multiple miniaturized completion can be described as miniature conventional systems bunched together.

The *tubingless completion* is an inexpensive option for shallow, low-pressure wells that produce only dry gas. They are also used on

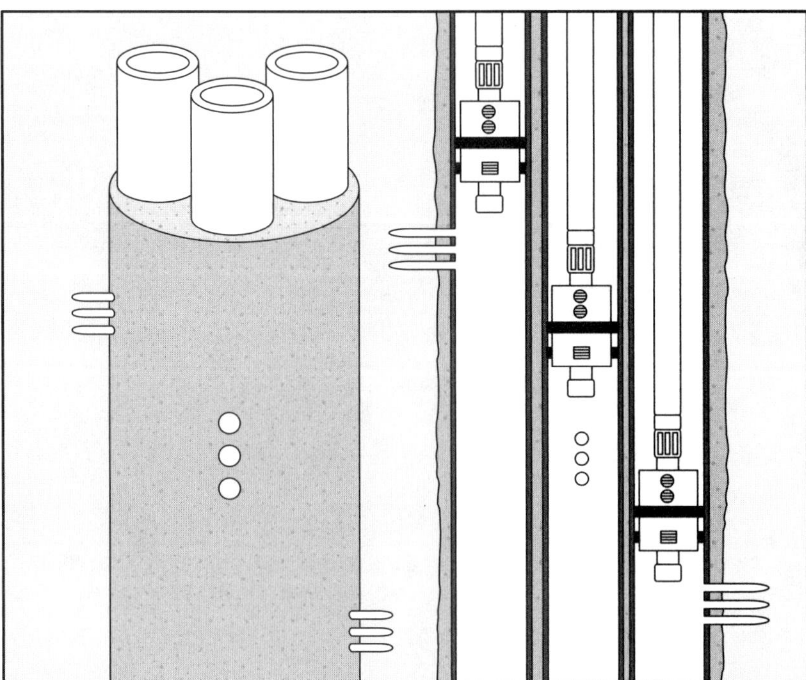

Figure 25. Miniaturized multiple completion

deeper wells when reserves are too low and economics too poor to justify conventional completions. Tubingless refers to the fact that no tubing is run in the production casing. The casing is sized according to pressure and flow velocity needs, is smaller than normal, and is often tubing. In multiple tubingless systems, two or more casing strings are cemented into a single hole, and oriented perforating techniques are used to match tubing to a particular producing zone and to be certain that perforations from one string do not penetrate the parallel string. Multiple tubingless configurations resemble multiple miniaturized designs except, of course, that in tubingless ones the casing has no downhole tools or tubing. Since no tools are needed in the casing and since the casing is often smaller than normal, both single and multiple tubingless completions usually are less costly than conventional completions.

Cost Versus Potential

Cost is an important factor in planning a well completion. Cost includes not only the well's price tag but also any losses of personnel, equipment, time, or money from a poor safety record, corrosion and erosion damage, excessive workovers, and environmental harm. Cost also includes the expected expenses of routine well service, production maintenance, workover, and recompletion.

A successful completion balances this cost with the well's potential to produce profitable amounts of hydrocarbons. The balance lies between the low-priced completion that falls short of the well's producing potential and the high-priced completion that wastes money on unneeded options and lowers the well's return on investment. The best completion costs no more than necessary to provide the well the chance to generate the maximum rate of return with the greatest level of safety. Overall efficiency, not a big budget, is the route to success. High cost does not always mean high profit potential, nor does low cost mean low potential. The investment for a high rate, short-term producer may be far less than for a low rate, long-term producer, and yet the two wells may produce the same number of barrels (cubic metres) of oil or cubic feet (cubic metres) of gas during their unequal lives. The cost effectiveness of the completion depends on the well itself—not on its price.

Drilling and completing a well is no different from any other business. It is important not to spend too much or too little. If a good balance between cost and potential can be found, then the well can be a success.

To summarize—

Completion design has three parts to the design process: a completion method most suitable for the well; a completion system to bring fluids to the wellhead; and a completion cost versus production potential.

Completion Methods

1. Open-hole completion is where casing is set and cemented near the top of the production interval and is only used in thick, continuous, consolidated carbonate reservoirs, horizontal wells, and coal bed completions.
2. Perforated completion—perforations are made at precise depths that provide wellbore communication at controlled pressure and flow rates.

Flow paths—most wells have either a single flow path or multiple paths that produce certain fluids at specific rates and conditions throughout a well's producing life.

Completion systems—an assembly of tools designed to meet specific needs for efficient production of the well.

Conventional systems—are cased, cemented, perforated completions with tubing and one or more packers, the most common conventional system being the single-zone completion with tubing and packer.

Specialized systems—two inexpensive, specialized systems are 1) smaller than normal miniaturized completions that use a small diameter hole to drill, and 2) tubingless completion; used on shallow, low-pressure wells that produce dry gas, where no tubing is run in the production casing.

Cost versus Potential

- Cost can include routine well service, production maintenance, workover, and recompletion, as well as any losses of personnel, equipment, time, or money from a poor safety record, corrosion and erosion damage, workovers, or environmental harm.

Completion Tools

Today's oil field technology offers numerous possibilities for completing a well. While every well is unique and for every well there is a unique completion, all completion designs include certain options. Although a range of modern completion equipment that seems almost limitless provides these options, all completion tools fall under a few basic classes. Each class fills a specific need in the completion design. The numerous combinations of these basic tools allow a completion tailored to suit the well. These basic tools include tubing, packers, seating nipples, flow control equipment, erosion control equipment, wellheads, and when necessary, artificial lift.

Tubing

A tubing string is an important completion option and is usually the first consideration in completion design (fig. 26). When used with various equipment options, tubing is the heart of a completion system.

Figure 26. Running tubing into a well

Tubing Strength Ratings

Every type of tubing is assigned American Petroleum Institute (API) ratings for tensile, collapse, and burst strength. *Tensile strength* is the maximum pull that the tubing can bear from its own weight plus any added pull required to unseat packers or anchors. If the pull is too great, the tubing will stretch until it parts. *Collapse strength* is the maximum bearable external differential pressure (the pressure in the casing annulus around the tubing minus the pressure inside the tubing). Annular pressure may be great enough to crush the tubing, pinching it like a drinking straw. As a result, costly recovery and replacement of the failed tubing is necessary. *Burst strength* is the maximum differential internal pressure that the tubing can withstand (the pressure inside the tubing minus the pressure in the casing annulus around the tubing). Pressures in wells that are shut in or undergoing stimulation treatment may become great enough to balloon the tubing. Still greater pressure may cause it to burst. Other joints of tubing may weaken by exceeding any forces greater than tubing design permits, in which the entire string could require replacement.

Tubing Connections

The tubing connection, or coupling, is a potential weak point in the tubing string. However, a well designed box and pin can be made up so that the connection is actually stronger than the tubing on either side of it. The design and thickness of the tubing connection compensates for the weakness caused by threads. Tubing connections vary widely in style and physical characteristics. Some connections are customized for special needs.

Tubing Movement

Tubing movement is a major consideration in completion design. Defined as a change in tubing length and also called breathing, tubing movement is caused by a number of things. Temperature variations in produced hot fluids or in hot or cold fluids, which are injected into the well, expand and contract the tubing. Pressure variations, both inside and outside the tubing string, balloon the tubing and shorten it or reverse-balloon it and lengthen it. Sucker rod pumping is the most familiar type of tubing movement. With each pump stroke, the transfer of fluid back and forth between the rod string and tubing can move the tubing several feet up or down. Such tubing movement can cause tubing failure, especially at pump rates of thousands of strokes per day. The rubbing of tubing against casing wears holes in both pipe strings. The repeated stretching and squeezing of the tubing may fatigue it until it breaks. Energy absorbed by tubing movement does not lift oil and water but increases investment and operating costs. To prevent or minimize tubing movement, a tubing anchor or packer may be set near the lower end of the tubing string to hold it in place.

Production Packers

Several hundred kinds of packers are available for use in acidizing, cementing, fracturing, gravel packing, testing, and thermal and waterflood injection, as well as in production applications for perforated and open-hole completions. A packer acts as a stopper in the production casing string. It provides a secure seal between everything above and below where it is set, allowing control of flow and pressure at that point. During production a packer is an optional but common completion tool that, combined with tubing, controls erosion, corrosion, pressure, fluid velocity, and heading in the well.

Erosion from sand and sediment in produced fluids can quickly damage casing. Just as sand pushed by water or wind can wear away rock, fine particles pushed by flowing crude can wear completely through metal pipe. A packer and tubing completion protects the casing above the packer from erosion, since formation fluids contact the casing only below the packer. The casing below the packer is usually protected externally by cement. Tubing, of course, may erode. But unlike cemented casing, it can be pulled and repaired or replaced.

Packer completions prevent casing corrosion in much the same way as they prevent erosion. Corrosion from carbon dioxide, hydrogen sulfide, and salt water can eat through casing and cause it to fail. The packer and tubing confine the corrodents and save the casing. Corroded tubing can be replaced, or corrosion-resistant tubing, which is cheaper than corrosion-resistant casing, may be used. Tubing, since it is smaller in diameter than casing, can withstand more pressure than casing. A packer diverts all pressure up the tubing, eliminating stress on the casing.

Since the well's pressure is diverted through the tubing, which is smaller in diameter than casing, flow energy is concentrated and more efficient. This efficiency helps to hold fluid velocity above the minimum speed needed to keep the well cleaned out. If an oilwell flows too slowly, sand and sediment settle and can clog the well. This clogging, sometimes called *sanding off*, can cost far less to prevent using packer and tubing than to correct with workover. A gravel-packed screen liner can help to prevent sand from entering the flow stream.

Loading, caused by liquids (condensate and water) forming as a gas well is being produced, can "log off" and cause the well to die and require swabbing. Smaller diameter tubing is sometimes installed to keep velocity of gas and liquids at a rate high enough to clear liquids from the tubing and prevent loading.

Figure 27. Packing elements

Figure 28. Slips

Heading is a type of poor flow in packerless, flowing oilwells. If pressure drops below the minimum needed to push oil to the surface, then the well stops flowing, gas comes out of solution, and pressure builds in the annulus. Accumulated pressure forces oil up the tubing, but gas from the annulus also escapes by the same route. Oil is produced slowly and in slugs, with halts in flow while gas pressure rebuilds. The use of a packer raises production efficiency because gas does not need to come out of solution to push oil up the tubing. Instead, dissolved gas works directly and immediately without the stop-and-start flow characteristic of heading.

Besides helping tubing to work to full advantage, packers are effective as a means for isolating producing zones from one another in the same wellbore. This practice is an important part of production management since neighboring formations do not usually behave alike, especially over long periods of time. A low-pressure zone may become a "thief zone" to a high-pressure zone, with fluids produced not to the surface but into the low-pressure zone as the two formation pressures try to equalize. Zone isolation prevents the problem of thief zones. Moreover, laws in many parts of the world require zone isolation.

Most packers share some basic components. All packers have *packing elements* (fig. 27) and *slips* (fig. 28). The packing elements, which are dense rubber, washer-shaped pieces sometimes separated and supported by horizontal metal disks, are designed to expand against the casing and seal off the annulus. Slips are toothed metal fingers that grip the inside of the casing and stop packer movement. The packing elements provide only a seal. If the packer moves while the packing elements are expanded, they will rub off against the casing and will eventually tear and fail. Upward pointing slips prevent downward packer movement; downward pointing slips prevent upward packer movement. Some packers have two sets of slips pointing toward each other to seat the packer tightly and prevent movement in either direction.

Slips are moved outward and held against the casing by a cone that is part of the packer setting mechanism. The setting mechanism allows the packer to telescope in upon itself. As the packer shortens during setting, the cone is forced behind the slips and the packing elements are expanded. A locking mechanism inside holds the packer in a shortened, set position. Packers that are designed to prevent movement in either direction are sometimes equipped with two cones.

To help the slips hold, some packers are equipped with hydraulic *hold-down buttons* (fig. 29). A hold-down button is a toothed disk with a hydraulic mechanism behind it. The mechanism uses reservoir pressure from below to press the buttons firmly against the casing. The greater the pressure inside the packer, the harder the buttons grip to prevent the packer from moving upward in the casing. Hydraulic hold-downs are also frequently used during stimulation operations, especially in shallow wells when the weight of the tubing is insufficient to prevent the packer and tubing from being pumped up and out of the hole by high pressure during stimulation.

Many packers require rotation to activate their setting mechanisms. These packers are equipped with *drag blocks* (fig. 30) or *drag springs* (fig. 31), also called friction blocks and friction springs. Drag blocks are spring loaded, rectangular buttons that depress easily as the packer is run in but stay out and catch on the casing when the packer is rotated. Drag springs, which look like small centralizers, perform the same role. While the upper part of the packer rotates for setting, drag blocks or springs keep the lower part of the packer from moving with the upper part.

Figure 29. Hold-down buttons

Figure 30. Drag blocks

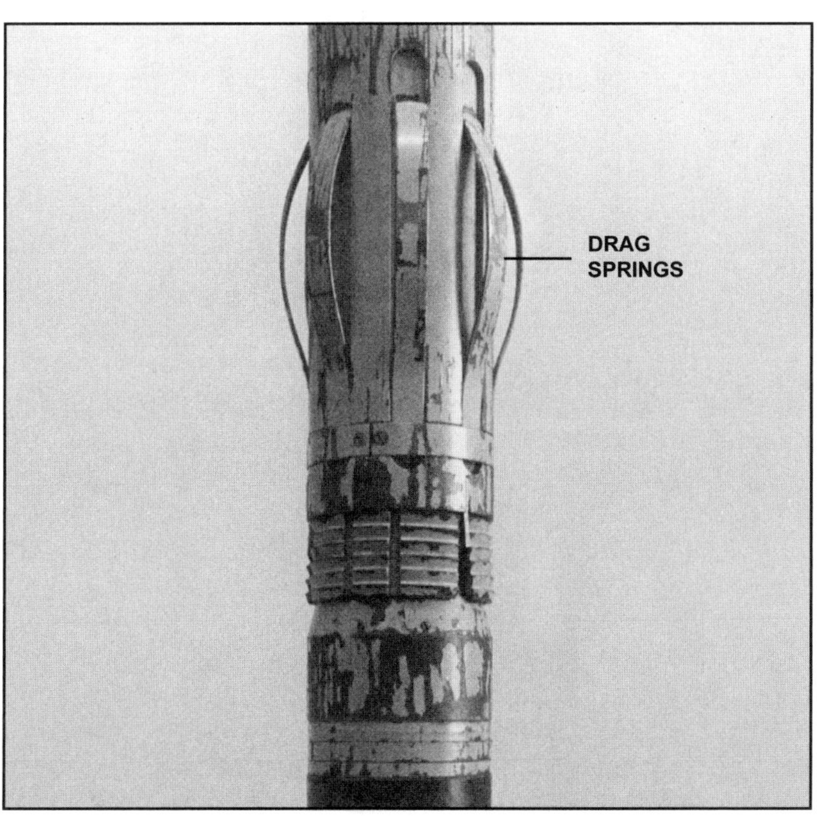

Figure 31. Drag springs above slips and packing elements

Some packers have threaded connections for tubing. Others have smooth seal bores designed to receive special tubing seal assemblies that are landed through the packers for production. Packers with threaded connections are usually retrievable; seal-bore packers are usually permanent.

Retrievable Packers

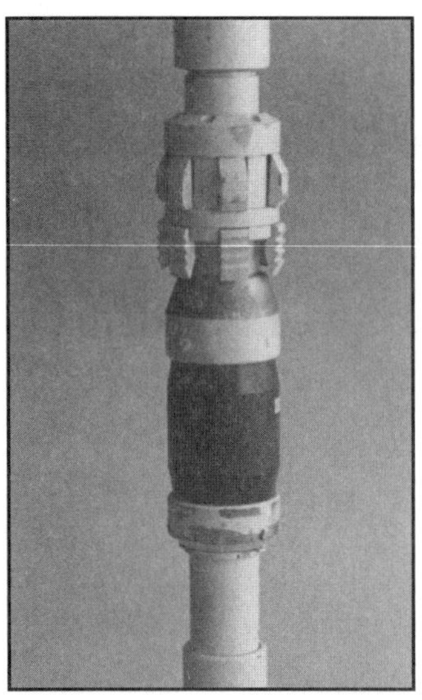

Figure 32. Tension packer

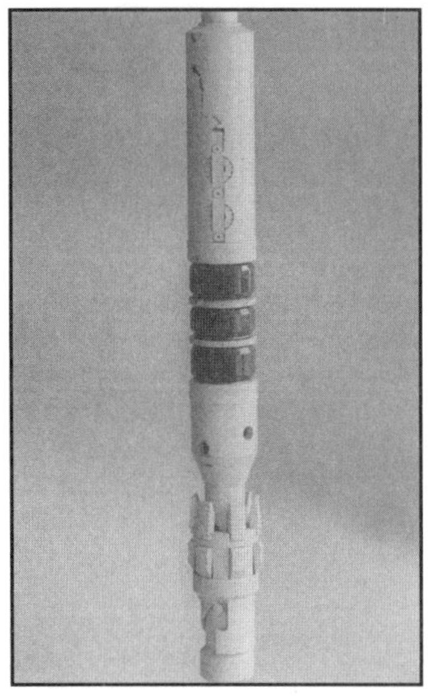

Figure 33. Compression packer

Retrievable packers may be set mechanically, hydraulically, or electrically. Mechanical-set packers are run in on drill pipe or tubing and require some type of movement to set them, often a combination of rotation and set-down weight. Some mechanical-set packers are locked in the set position with tubing tension or compression. Tension packers (fig. 32) require a specific upward pull, while compression packers (fig. 33) require a designed minimum set-down weight. Packers that need neither tension nor compression to remain set are sometimes called neutral-set packers.

Hydraulic-set packers are run on tubing or wireline. Some types contain piston chambers that compress under formation pressure to set the packer. Other types are activated by pressure from pumps at the surface. Varieties of setting mechanisms are designed to respond to hydraulic pressure.

Retrievable packers set by electric line are the least common. They use a small explosive charge to set the slips and expand the packing elements. Since they may be set on wireline, electric line-set packers can be placed at very precise depths when required.

When necessary, retrievable packers are easily moved or removed. Straight upward pull or rotation, or both, usually release them. Their retrievability allows them to be moved a few feet and reset, or moved many feet, or pulled completely, as during workover. Because of their intricate setting and release mechanisms, retrievable packers are larger, more expensive, and not as strong as permanent packers are.

Permanent Packers

Permanent packers (fig. 34) may be set by any of the methods used to set retrievable packers but, unlike the retrievables, are most commonly set electrically. Permanent packers are neutral and are very secure against upward or downward movement. They have two sets of slips, need no hold-down buttons, and are simple and reliable. Their basic drawback is that they must be drilled out for removal, so they are often made of brass, cast iron, or other easily drillable materials. Semipermanent packers have been developed that have the strength of their permanent versions and the advantage of release and retrieval, but once released they cannot be reset and must be replaced. Semipermanent packers can be retrieved using a special tool run in the hole on tubing.

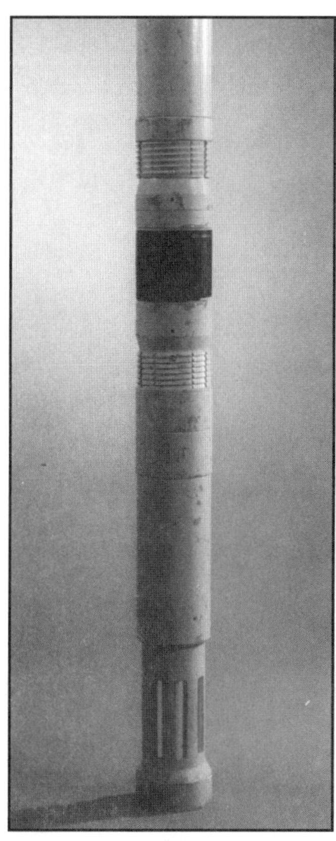

Figure 34. Permanent packer

Most permanent packers use seal-bore assemblies to provide confinement of well fluids to the tubing string. They have smooth bores designed to accommodate a tubing seal assembly, which is made up and run on a tubing string. The tubing seal assembly is encircled with many flexible rings to provide a tubing-to-packer seal while allowing some freedom of movement for tubing under the stresses of pressure and temperature changes. The tubing and tubing seal assembly can be easily pulled while the packer remains in the hole. Another similar tool is the polished bore receptacle (PBR). The packer of choice in deep high-pressure gas wells equipped with a production liner run below intermediate casing is the PBR with a tubing seal assembly. The PBR is run as an integral part of the production liner. The seal assembly is run on tubing and strung into the PBR. Because deep wells typically exhibit high temperatures, a PBR of several feet (metres) in length and a long seal assembly can be run to compensate for lengthening and shortening of the tubing during temperature changes.

Most packers are single packers. That is, they provide a seal for one string of tubing. If more than one zone is to be produced from a wellbore, then a dual packer that accommodates two tubing strings may be used. Though rare, triple and quadruple packers may be used to complete wells drilled through several pay zones.

Seating or Landing Nipples

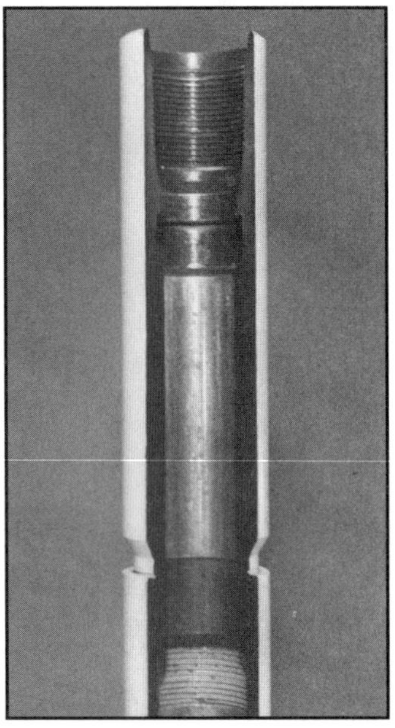

Figure 35. Seating nipple

A tubing string may be run without a packer, but it is almost always run with a seating, or landing, nipple. A seating nipple is a threaded length of pipe with a smooth, polished bore that may contain one or more special horizontal grooves or slots (fig. 35). The locking grooves are designed to receive wireline tools such as standing valves (a type of check valve) and plugs. The grooves and locks of the wireline tools are of certain sizes and shapes so that they must match the grooves and slots in the nipple.

Seating nipples are made up on the tubing string wherever wireline tools may be needed, whether at the time of the initial completion or later in the well's life. Some types of flow control equipment, for instance, are run on wireline and are set in nipples and become virtually permanent. Seating nipples are also sometimes called landing nipples.

Another type of seating nipple is constructed with a smooth bore internal diameter slightly less than the inside diameter of the tubing. This provides two very important functions. The first is to prevent swabs and similar tools from being run past the bottom of the tubing or from falling out the end of the tubing in the event the swab line or the braided line breaks during completion, maintenance, workover, or recompletion. Pulling tubing to retrieve a swab left in the hole by a line failure is almost always less costly and time consuming than attempting to fish a swab lost in casing. The second use is as a landing nipple for rod pumps or tubing stop for plunger lift.

Flow Control Equipment

Flow control includes the planning of flow paths above and below the packer through both the tubing and the annulus. It also involves safeguards against flow pressures above maximum levels. Types of flow control equipment commonly installed during initial completion include sliding sleeves, check valves, and subsurface safety valves.

Sliding Sleeves

Sliding sleeves are made up on the tubing string at points where periodic or continuous communication between the tubing and casing-tubing annulus is needed (fig. 36). A shifting tool run on wireline is used to slide the sleeve open or closed. The open sleeve allows circulation either way: down the tubing and up the annulus or down the annulus and up the tubing. A sliding sleeve above a packer, for example, permits changing of fluids placed in the annulus over the packer for pressure or corrosion control.

A sliding sleeve between two packers used to isolate a pay zone may be opened or closed to produce that zone selectively or to commingle production from both zones.

Check Valves

Check valves and standing valves are one-way valves installed in seating nipples (fig. 37). Some allow flow up the tubing but not down it. This type of check valve is often placed beneath hydraulic tools, such as packers, which are activated by pumping down the tubing string. Another type of check valve is designed to resist downward flow only until a certain static pressure is reached, above which fluid may be pumped downward through the valve. Still other check valves allow only downward flow, as in injection wells. Check valves are also called check plugs.

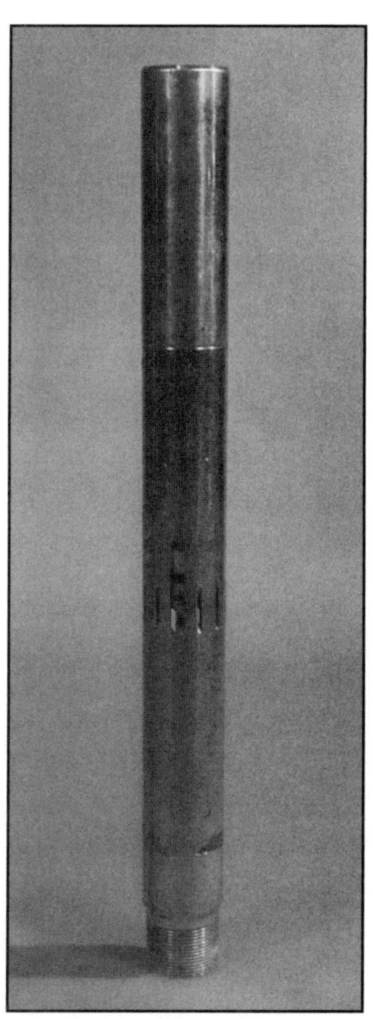

Figure 36. Sliding sleeve

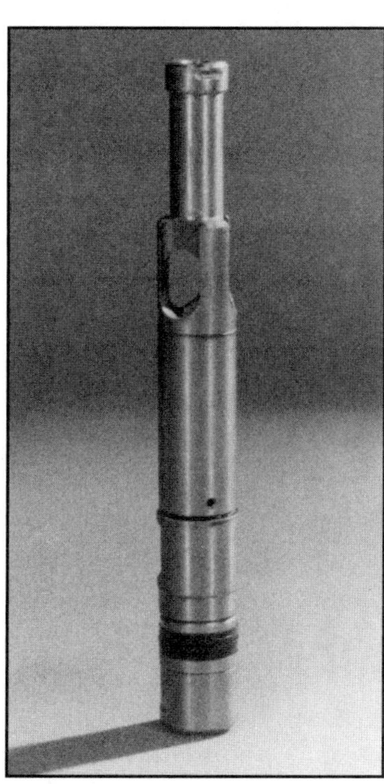

Figure 37. Check valve

Subsurface Safety Valves

Subsurface safety valves shut in the well should surface systems become damaged or are removed. They are sometimes called storm chokes on offshore wells (fig. 38). For instance, if a wellhead is knocked off its well, nothing may prevent a dangerous and costly blowout except the subsurface safety system. Besides

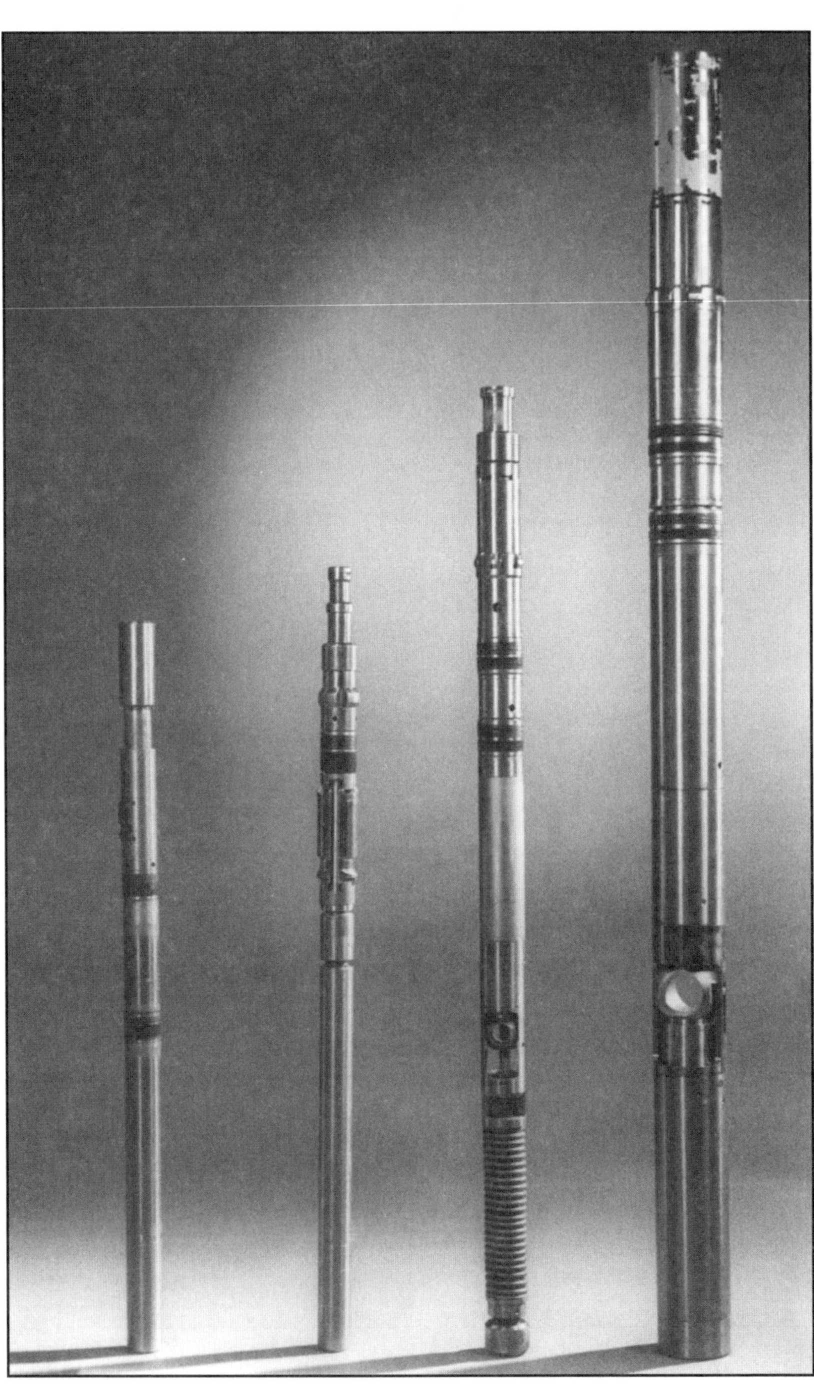

Figure 38. Subsurface safety valves

pressure, some subsurface safety valves also respond to excessive flow velocities or temperatures. A well may have more than one valve for extra security.

Subsurface safety valves are made up on tubing or set in seating nipples by wireline. Both tubing-mounted and wireline-retrievable types may be subsurface controlled or surface controlled. Subsurface control valves respond to flow rates above preset maximums by shutting in immediately. Surface control valves have surface sensors that detect sudden increases in pressure or temperature and bleed off fluid from a hydraulic control line to close the safety valve downhole. Since erosion, corrosion, or frequent adjustment may damage the valves, wireline-retrievable types are sometimes preferred because they are easier to remove and check or repair. On the other hand, tubing-mounted types are generally larger and therefore more resistant to damage and wear.

Erosion Control Equipment

The flow of formation fluids toward the wellbore may loosen and carry along many small particles of rock. This sandy sediment moves forcefully and can wear through metal completion tools. Erosion occurs where flow velocity is highest, where flow paths change direction, or where fluid passages suddenly narrow. A change in flow-path direction occurs when fluids strike tubing as they blast through perforations into the wellbore. Sudden narrowing of the fluid passage occurs at any nipple or wireline tool that varies the inner diameter of the tubing string for a short distance of a few inches (millimetres) or a few feet (metres). Blast joints control external tubing erosion while flow couplings control internal tubing erosion.

Blast Joints

A blast joint is a thick tubing sub designed especially to withstand erosion. The effect of oil-borne or gas-borne sediment on tubing just inches away from perforations is similar to that of sandblasting. Blast joints are often sheathed in tungsten carbide, an extremely hard material many times more wear resistant than steel. To protect vulnerable sections of the tubing string, blast joints are made up on the tubing wherever it hangs opposite casing perforations.

Flow Couplings

A flow coupling, in a sense, is a blast joint turned inside out. Its function is to reduce turbulence and resist wear from fluids traveling through the tubing string. Flow couplings are approximately 2 to 3 feet (0.5 to 1 metre) long, made of high strength, carbon alloy steel, and have smooth, polished bores. They are generally made up just above and just below each seating nipple in the tubing, as well as on either side of sliding sleeves (fig. 39). Screen liners and gravel packs may also be thought of as erosion control means, but rather than resist sand abrasion, these tools prevent erosion by keeping sand out of the produced fluids.

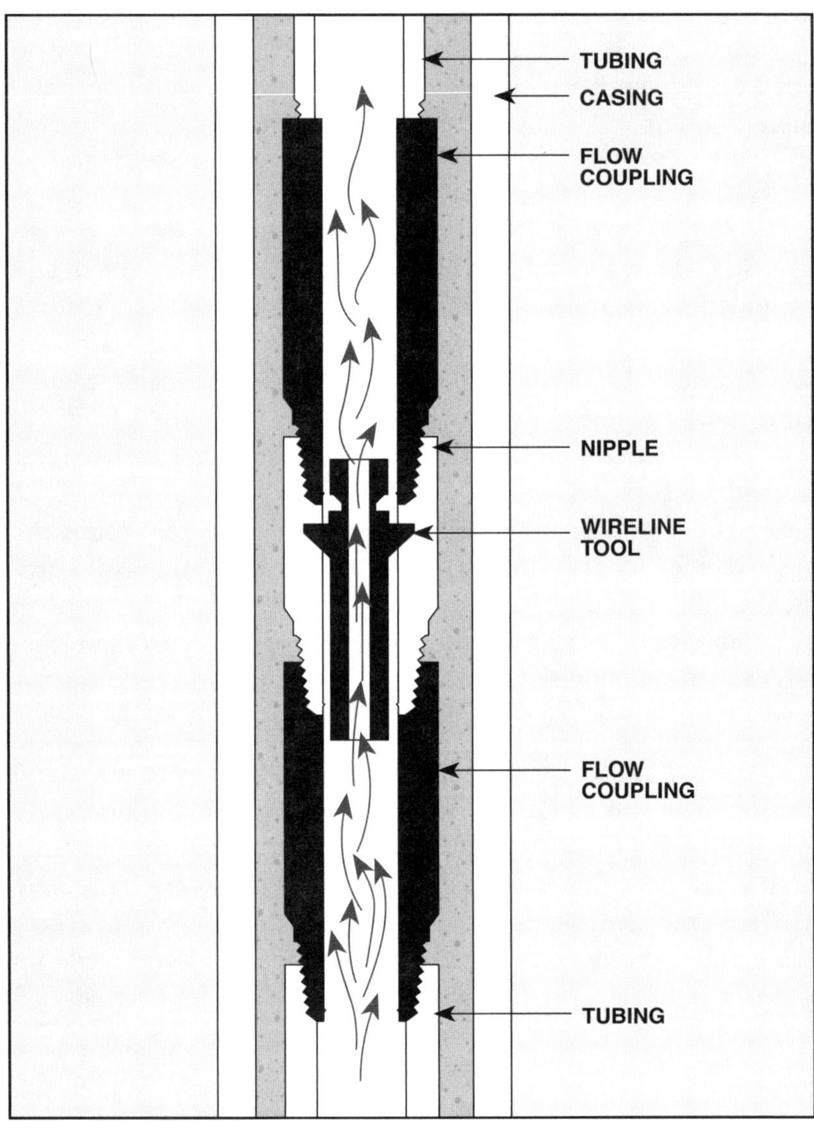

Figure 39. Flow coupling

Wellheads

The wellhead includes all equipment placed on top of the well to support tubulars, provide seals, and control the paths and flow rates of fluids. Wellheads include at least one casinghead and casing hanger, a tubing head and tubing hanger, and a Christmas tree (fig. 40).

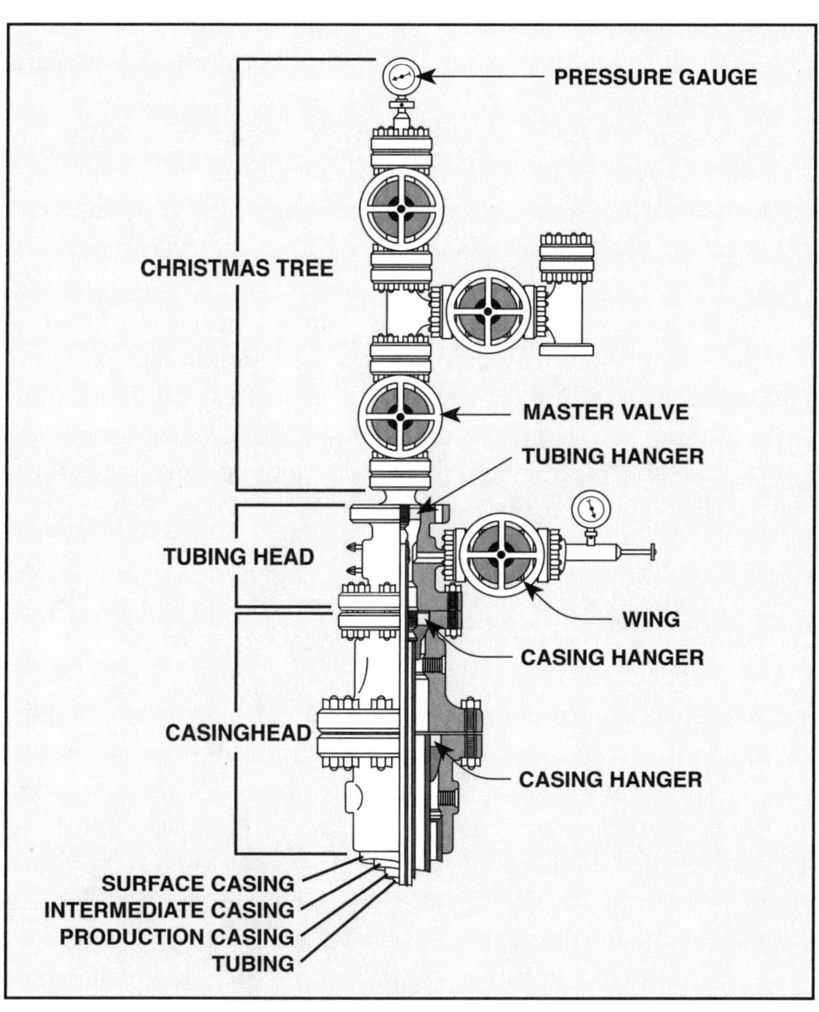

Figure 40. Wellhead components

Casingheads

Casingheads are attached to surface casing or to another casinghead to provide a hanging point for the next string of casing. If there is one casinghead, it is usually welded to the surface casing (in some low-pressure applications it is screwed onto the surface casing), and the production casing is hung from it. If more than one casing string is used inside the surface casing, then more than one casinghead may be needed. An intermediate casinghead may be added with each new casing string until the production casing has been hung.

The top of a casinghead has a cone-shaped bowl that holds the *casing hanger*. A casing hanger is a set of slips that grips and supports a casing string. Casing is suspended from the hanger, rather than from the casinghead itself. Metal and rubber packing rings fit over the slips to complete the casing hanger assembly and provide an annular seal. Threaded or flanged outlets on the sides of the casinghead allow access to the sealed annulus for pressure gauges that warn of packer, tubing, and casing leaks and monitors for wells equipped with artificial-lift systems such as beam pumping, submersible pumping, hydraulic pumping, progressing cavity pumping, and plunger lift.

Tubing Heads

The tubing head is attached to the uppermost casinghead. It may also be attached to the production casing. Its *tubing hanger* seals the annulus between the tubing and the production casing and supports the weight of the tubing string. The tubing hanger is carefully designed not to crush or deform the pipe it holds. Outlets on the sides of the tubing head allow access to the production annulus for gauging pressure or producing fluids. Another type of tubing hanger uses matched sets of flanges to support the tubing. The tubing is screwed into the upper companion flange, which is then attached to the lower companion flange using nuts and bolts.

Casingheads and tubing heads may be threaded or flanged. Since threaded heads are not as strong as flanged ones, they are used mostly on shallow, low-pressure wells. Flanged heads, which are also called spools because of their shape, are attached by bolting, although the first casinghead is often welded to the surface casing. The tubing head may also have anchor screws to provide extra tubing support. Casingheads and tubing heads are rated by maximum working pressure, which may range from 1,000 psig (7,000 kilopascals) to more than 20,000 psig (140,000 kilopascals).

Christmas Trees

The Christmas tree includes everything over the tubing head (fig. 41). It is composed of valves, gauges, and pipe that regulate, measure, and direct flow from the well. The main valve of the Christmas tree is the master valve, which is attached above the tubing head and on which the rest of the tree is mounted. The master valve is used to shut off production and is important in well service and workover.

Wellheads for multiple completions are basically the same as for single completions. When a second zone is produced up the annulus, pipes from the side of the tubing head, called wings, join flow to the Christmas tree and provide pressure gauge hookups. In multiple string designs, the casinghead and tubing head may be larger than normal to make room for more than one string of tubing, and the tubing hanger suspends more than one string. Multiple completion Christmas trees often have more than one master valve and are generally more complex than single trees.

Figure 41. Two types of Christmas trees

Some production methods, such as artificial lift, affect the choice of completion tools. In areas where low-pressure oilwells are common, sucker rod pumping is a frequent artificial-lift method. The beam pumping unit that moves the sucker rods is the most obvious difference in this type of completion (fig. 42). Other changes, such as the substitution of a pumping wellhead to accommodate the sucker rod string, are also needed. Special valves used particularly for pumping are installed at the surface. Other types of valves and tools, such as gas-lift mandrels, are needed for gas lift, which is an artificial-lift method where produced gas is injected back into the well to help produce liquids. Hydraulic lift systems, which use high-pressure oil or treated water to reciprocate a downhole pump, require a still different surface hookup. Progressing cavity pumping uses sucker rods that are rotated, instead of reciprocated as in beam pumping. Progressing cavity systems are especially useful for high rates of production from shallow wells, for lifting high viscosity crude oil, and for pumping oilwells that produce substantial amounts of sand along with the crude. Artificial-lift equipment of one kind or another is added to almost all oilwells, whether initially or later in the well's life.

Figure 42. Beam pumping unit

Completion Procedures

Almost all modern wells except coal bed methane wells and some extended reach horizontal wells are cased and cemented. They are usually equipped with tubing and packer and are perforated and are often stimulated. Therefore, it is useful to understand the basic ideas and techniques of perforating.

To summarize—

Completion tools—are made up of several combinations of tools that include tubing, packers, seating nipples, flow control equipment, erosion control equipment, wellheads, and artificial lift.

Tubing

- A tubing string is an important completion option and is usually the first consideration in completion design. When used with various equipment options, tubing is the heart of a completion system.

Tubing strength ratings

- Tensile strength rating is the maximum pull that tubing can bear from its own weight plus any added pull required to unseat packers or anchors.
- Collapse strength rating is the maximum bearable external pressure minus the internal pressure.
- Burst strength rating is the maximum internal pressure that the tubing can withstand minus the external pressure.
- Tubing connection is a potential weak point in the tubing string.
- Tubing movement is a change in tubing length (also called breathing) caused by temperature and pressure variations or sucker rod pumping.
- Production packers are used for acidizing, cementing, fracturing, gravel packing, testing, thermal and waterflood injection, as well as production applications for perforated and open-hole completions.
- Loading is caused by liquids forming as a gas well is being produced, and can cause the well to die and require swabbing.

- Retrievable packers may be set mechanically, hydraulically, or electrically, and can easily be moved or removed. Hydraulic set packers are run on tubing or wireline.
- Permanent packers may be set by any of the methods used to set retrievable packers but, unlike the retrievables, are most commonly set electrically.
- Semipermanent packers have strength and the advantage of release and retrieval, but once released cannot be reset and must be replaced.
- Seating or landing nipples are threaded lengths of pipe with a smooth, polished bore that may contain one or more special horizontal grooves or slots.

Flow Control Equipment

- Common types of flow control equipment are sliding sleeves, check valves, and subsurface safety valves.
- Blast joints are thick tubing subs designed to withstand erosion, often sheathed in tungsten carbide.
- Flow coupling is a blast joint turned inside out used to reduce turbulence and resist wear from fluids traveling through the tubing string.
- Wellheads include at least one casinghead and casing hanger, a tubing head and tubing hanger, and a Christmas tree.

Perforating

The advent of gun perforating in the 1930s made it easy to shoot round, uniform holes in cemented casing. Until then, passages for formation fluids had to be made in cased, cemented wells with clumsy, unreliable mechanical perforators that slowly gouged each hole needed, sometimes badly damaging the casing in the process. The disadvantages of early perforating techniques made the open hole a common completion choice at the time.

Gun perforating radically changed completion technology. It allowed the entire well to be cased through and specific zones to be produced by means of accurately placed perforations. Acidizing and fracturing techniques almost always work better in wells where the petroleum reservoir is isolated from other formations. Also, multiple zones separated from one another in the same wellbore produce more efficiently. Zone isolation has become a legal requirement in many areas with commingling allowed only after proper request and permission have been granted. Effective perforating techniques have helped to make all these things possible.

Originally, gun perforating used actual bullets fired at the casing to penetrate cement and formation rock. It was cheap, effective, and fast. But bullet perforating has pressure and temperature limitations. During World War II, bazooka technology was used to develop jet perforating, which has since become the method of choice for a very high percentage of cased-through completions.

Jet Perforating

Jet perforating is the most versatile perforating method available today. Neither pressure nor well fluids affect its performance and, where formation temperatures exceed 350° Fahrenheit (175° Celsius), special explosive charges may be used to prevent uncontrolled firing from heat. Also, in dense, highly compressed rock, jet perforating performs better than bullet methods because the depth of penetration of jet perforations is not shortened so drastically with increases in formation depth as is bullet perforating.

Jet perforating, like the bazooka, works on the principle of propelling a high-energy stream of fine particles into a material to make a hole in it (fig. 43). In a jet perforating gun, a cone-shaped charge explodes and generates a powerful shock wave. The shock wave exerts up to 6,000,000 psig (40,000 megapascals) on a metal liner. The liner is also cone-shaped since that form improves its breakup into a jet particle stream. Rather than destroying casing, cement, and formation rock, the jet stream pushes them aside as it moves through. Traveling at more than 20,000 mph (over 32,000 kilometres per hour), the jet may reach as far as 52 inches (130 centimetres) in fractions of a second before running out of energy.

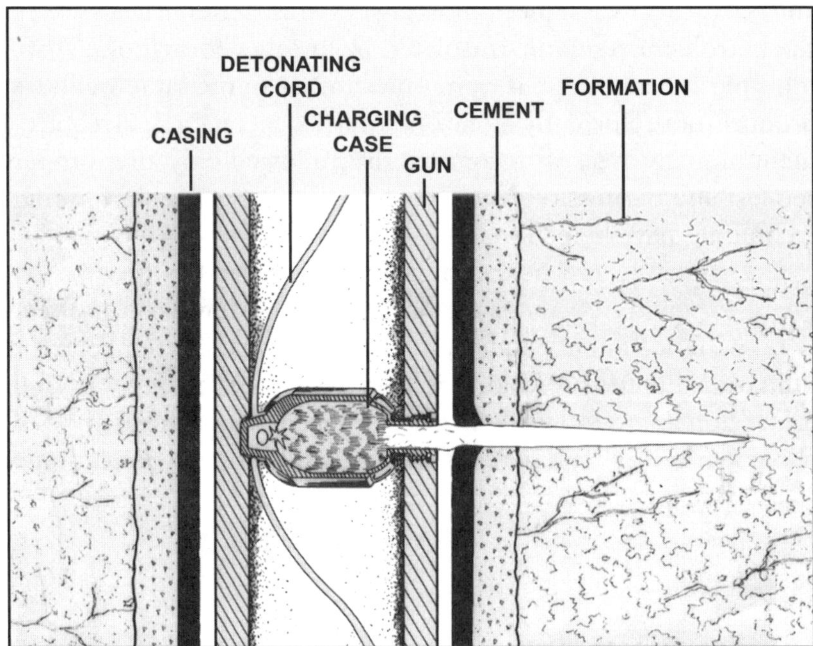

Figure 43. Jet perforating

The shaped charge is held in a metal case, open on one end, (fig. 44). In the closed end is a detonator, which is set off by a detonating cord that also runs through other charges in the gun. The explosive is usually cyclonite (cyclotrimethylenetrinitramine, also called RDX) and the cone liner is either copper or copper-zinc alloy. A related explosive, HMX, is used in wells where temperatures are as high as 425° Fahrenheit (over 200° Celsius).

Figure 44. Shaped perforating charge case

Electric Line Conveyed Perforating Guns

The choice of perforating gun depends basically on the diameter and length of the perforation desired, the size of casing and cement sheath to be penetrated, and on whether the gun is to be lowered through casing on conductor line, on tubing, or through tubing on conductor line. Service companies offer a variety of perforating guns to carry shaped charges.

The bigger the perforations needed, the stronger are the jets required to make them, the bigger are the cyclonite charges required, and the bigger is the gun required to hold the charges. Metal tubes called retrievable guns can generally fire the strongest charges. If the perforating gun is to be lowered through tubing, which is smaller than production casing, then the gun will necessarily have to be smaller, too. The through-tubing gun has a smaller, lighter steel tube than the retrievable casing gun. It is designed to run through the bottom open end of the tubing string to perforate casing in the lower zones of the well. Semiexpendable and expendable guns, made of lightweight aluminum links, usually have the smallest outer diameters and are commonly used when small-OD guns are needed. Some of these small guns are partially or completely destroyed when the jet charges are fired.

Perforating Procedures

Once the perforating charges and gun have been chosen and the thickness of the zone to be perforated has been decided, the perforating service company assembles the gun or guns and brings them to the well. Expendable guns can be assembled quickly and efficiently on location. Several guns may be rigged together for perforating thick zones.

A completion fluid may be circulated into the well to displace drilling mud and clean the hole of debris that might plug perforations. Completion fluids are generally low solids brines, but they may also be oil-base fluid for the perforating of water-sensitive formations. Multiple volumes and types of chemicals are added to water to create a solution that will guard against damaging the formation from clay swelling, emulsion formation, and changing the rock's permeability to oil, gas, and water.

After the completion fluid (if used) has been placed in the well, the gun is lowered to the proper depth and fired. Some guns allow multiple firings for perforating more than one section of the well. Once perforating is finished the gun is brought back up or, in the case of expendable guns, the conductor line is simply reeled in.

Natural well flow can often clean perforations of crushed rock, mud, and charge debris so that the formation can produce to the wellbore. If not, several perforation cleanup methods may help. These techniques include washing with acid or other chemicals to dissolve and liquefy debris and back surging (suddenly flooding empty tubing) to remove debris from the perforations. Negative-pressure perforating, also called reverse-pressure perforating or perforating underbalanced, is done with wellbore pressure lower than formation pressure to provide a cleansing pressure surge through the perforations the moment they are shot. In wells where pressure control is not difficult, operators frequently swab some of the fluid out of the casing prior to perforating. This action sets up a pressure differential from the reservoir into the wellbore. When the well is perforated, reservoir pressure helps push materials in the damaged area near the well from the rock into the well. The opposite procedure is often used in overpressured reservoirs. Pressure is applied to the tubing and a through-tubing gun is lowered to the proper perforating depth. Pressure is maintained on the tubing until the gun is retrieved. This method is used to keep high-pressure fluid from entering the well and causing damage along with control problems until the perforating gun can be retrieved and the well is secured.

Tubing Conveyed Perforating Guns

Another entirely different method of locating perforations at the right place and maintaining well control involves the use of tubing conveyed perforating guns. Guns are assembled at the location or are preassembled and delivered to the location. As the name indicates, guns are lowered into the hole on the production tubing, properly located either by careful measurement or by correlation using a through-tubing gamma ray log, and the packer is set. Tubing may be partially swabbed to provide a positive pressure differential into the well from the reservoir and a bar is dropped to trigger the gun. Another application for the opposite condition—high-pressure gas wells—involves the same installation procedure but the gun is equipped with a preset firing device that is activated by pump pressure applied down the tubing.

Hydraulic Jet Perforating

A rarely used perforating method involves running an orifice-jetting device on the tubing to the desired depth. A pump truck then pumps sand-laden liquid down the tubing and through the jet where perforations are created by sandblasting. If desired, the tubing can be rotated back and forth and a slot cut through the casing and cement sheath into the formation.

Stimulation

As the oil industry has matured, more and more completions have been attempted in reservoirs that would not have been considered for completion 50 years ago. Also, many wells that are commercial producers naturally are stimulated to increase productivity and, consequently, rate of return. The high degree of success achieved has been a result of improvements in stimulation techniques, materials, and equipment. Hydraulic fracturing with the addition of sand proppants, having been frowned upon or possibly illegal, became an important procedure in the early 1950s to enable commercial completion of otherwise noncommercial reservoirs, and improved performance from commercial reservoirs. Carrying fluids can be classified as follows: (1) water based; (2) oil based; and (3) energized. Energized systems are most often water based and occasionally oil based.

Water Based

Carrying fluids usually are mixed and prepared in steel or plastic lined earthen pits at the well site or by adding gelling agents during pumping operations. They are usually of three subtypes: (a) water-base gel using various gelling agents and various additives to protect sandstone reservoirs, (b) water-base acid to perform massive acid fracs of carbonate reservoirs, and (c) water fracs which consist of pumping large volumes, often in excess of 10,000 barrels (1,590 cubic metres) of fresh water, into naturally fractured carbonate rocks. These rocks usually are partially composed of gypsum (anhydrite), which is washed out and dissolved by the fresh water.

Of the three basic types of stimulation systems, water-base gel is the only one which should be considered to be a true sand fracture treatment where the attempt is made to generate an artificial fracture and to prop it open with well-rounded, consistent sized sand grains to provide maximum permeability. Sand sizes range from medium to large in varying quantities from one-half pound per gallon (60 kilograms per cubic metre) to 12 pounds per gallon (1,500 kilograms per cubic metre) in true sand fractures. Sand is added in stages in increasing concentrations to provide maximum sand placement and optimum permeability.

Small quantities of very small mesh sand are sometimes added to water-based acid and water fracs to provide scouring of the fracture faces. Water-based acid sometimes includes small- to medium-sized sand to provide proppants for the fractures created by the process. This procedure may or may not be beneficial depending upon the crushing tendency of the formation rock.

Acid concentrations often are 22 percent to 28 percent strength to provide maximum dissolving of the carbonate rock. Fresh water, in volumes of one to three times the acid volumes, is pumped between the various acid stages to dissolve the sludge, which is a result of acid reacting with carbonate. The most commonly used material is hydrochloric acid because of its effectiveness and low cost. Corrosion inhibitors, demulsifiers, and clay stabilizing chemicals are also often added to the system.

Oil Based

Oil-based fracturing systems are used almost exclusively to transport sand into fractures in oil productive sandstone reservoirs. Amounts of sand are usually the same as water-based gels. Oil-based treatments are being phased out by water-based treatments for reasons of safety, cost, effectiveness, and versatility.

Energized Systems

Energized systems add nitrogen, carbon dioxide, or both, to water-based and oil-based systems to provide for more rapid cleanup of frac fluid and occasionally to improve sand carrying characteristics.

Any of the fracture systems can be pumped through the casing before installing tubing, or through the tubing, depending on circumstances and operator preference.

Running the Completion

Although every well has its own traits and poses a unique set of problems, the final steps of completion are similar for all wells. Overall considerations such as safety also change little. Completing a cased, cemented, and perforated well with packer and tubing follows a general pattern. Once the completion tools have been brought to the well site, the completion can begin.

Conditioning the Hole

To ready the hole for completion, the wellbore fluid in it must be prepared. This fluid may be drilling mud if a high-density fluid is needed for well control. Otherwise, a completion fluid may be substituted for the drilling mud. Displacement of the drilling mud helps to flush out mud cake and light debris such as rocks and pebbles and pieces of metal. Common practice, to avoid the cost of displacing and hauling off drilling mud, is to simply displace the cement wiper plug using completion fluid. Then the well is ready to perforate as soon as the cement bond log has shown that the primary cement job is satisfactory.

If the drilling mud is left in the hole, then it is circulated at a high rate. The rapid, turbulent circulation serves to remove thick mud cake that may have formed on the inner casing wall, since the hole may not have been circulated for some time. If the mud cake were left in the well, it might later fall down the hole and collect on top of the packer, creating a corrosive environment as well as hindering the packer's operation. Debris is removed from the circulating fluids.

A junk basket with casing scraper may be run on wireline or pipe to pick up junk, such as bit cones, slips, BOP stripper rubbers, pieces of cement and wood, or other foreign object that is too large or too heavy to be circulated out and that might damage completion tools or prevent their proper operation.

At some time during completion, a gauge ring may be run. The gauge ring verifies a full-diameter wellbore with no bent or collapsed casing sections that can affect the completion. Gauge rings or casing scrapers are usually run on pipe to make it possible to force the tool through irregularities and help to correct them. If a gauge ring or scraper is run after perforating, burrs and rough spots that were caused by perforations may also be smoothed out.

Inspecting the Tools

While the hole is being conditioned, the completion tools are unloaded and arranged in order of assembly. All tools are checked one last time for proper size and fit and for damaged or frozen parts. The completion expert and the unit operator discuss the sequence of events and other important points. These points include the maximum run-in speed and precautions against either sudden starts or stops or tubing rotation during run-in. Maximum run-in speed is particularly important, since lowering the packer too fast can cause a pressure differential high enough to destroy the packing elements. The tubing should be tallied and inspected visually after unloading.

Running and Setting the Tools

Depending on the type of packer used, the packer may be run and set before the tubing, or the two may be run and set together in one trip. If more than one packer is used, the completion may be installed in several sections requiring several trips. On some high-pressure completions that require high-strength premium equipment, tubing connections are pressure tested as tubing is run in the hole. Some types of packers are also pressure tested just before they are run.

Perforating

Some wells are perforated before the tubing string is run and set. Others are perforated afterward by lowering the perforating gun through the bottom end of the tubing or by running the gun as an integral part of the tubing. The time of perforation may change some aspects of the completion, but the overall process is the same.

Testing the Seated Packer

The method used to check whether the slips are properly set varies with the type of packer used. Weight is set down on a compression packer to verify that it does not slide downward. For tension packers, if added upstrain pull produces added tension on the tubing, then the packer is well seated. Neutral packers may be tested by set-down weight or upstrain pull. Permanent seal-bore packers set by wireline are also tested either way, according to the type of tubing seal assembly used, after the assembly is landed in the packer.

The packing element seal against the casing is tested on all packers in one of two ways. If the casing below the packer is perforated, then the annulus above the packer is pressure tested. If the casing is not yet perforated, posing no danger of formation breakdown, then pressure may be applied through the tubing to test for tubing or packer leaks.

Placing the Packer Fluid

Packer fluid may be placed in the casing-tubing annulus above the packer. The packer fluid displaces any fluid already above the packer. Sometimes the same completion fluid serves as perforating, completion, and packer fluid. The packer fluid must be low in solid materials because solids tend to settle over the packer and bake on at downhole temperatures. This occurrence prevents easy tool operation and removal. Packer fluids are usually brines, but they may also be freshwater or oil-base solutions. It often includes corrosion inhibitors and may be weighted with properly treated chemicals to assist in well control on future workovers and recompletions.

Landing the Wellhead

Casingheads have been attached as necessary throughout drilling. The blowout preventers are reinstalled as each new string of casing is run. By the time completion is begun, a well may have most of its wellhead already in place. To attach any remaining casingheads and the rest of the wellhead equipment, the BOP stack is unbolted and lifted off the casing. Casingheads and casing hangers are added, often only until enough of the wellhead is in place to control the well. Then the drilling contractor may remove the drilling rig to allow a completion rig to be brought in to install the tubing and packer, the tubing head, and the Christmas tree, and to assist during perforating and stimulation.

Cleaning up the Well

When the complete wellhead is in place, it is pressure-tested and checked for proper valve operation. To clean up after completion, high-pressure wells should never be flowed at rates higher than normal for production because of the danger of damaging the well, the equipment, and the reservoir. Fluids are flowed briefly into a pit or flared.

Flanging Up

The production behavior of the well is tested and, if economic production is obtained, it is at last tied to a gathering system or tank battery and put into production. High-pressure, high-deliverability gas wells are frequently connected to gas gathering lines, which had been installed during drilling. These installations are normally limited to development areas with nearby main gas lines and for wells that have shown high probability of being productive during drilling.

To summarize—

Perforating—jet perforating is the most versatile perforating method available today, performing better than other methods in dense, highly compressed rock.

Stimulation is made up of three stages:

(a) Water-based gel that uses various gelling agents and additives to protect sandstone reservoirs;

(b) Water-based acid that performs massive acid fracs of carbonate reservoirs; and

(c) Water fracs that pump large volumes into naturally fractured carbonate rocks.

Running the Completion

- To ready the hole for completion, drilling mud or completion fluid must be prepared.
- Depending on the type of packer used, a packer may be run and set before tubing, or both may be run and set together in one trip; if more than one packer is used, the completion may require several trips.
- Some wells are perforated before the tubing string is run and set and others are perforated afterward.
- Testing the seated packer is the method used to check whether the slips are properly set and varies with the type of packer used.
- Packer fluid may be placed above the packer in the casing-tubing annulus, which displaces any fluid already above the packer.

Glossary

A

acidize *v*: to treat oil-bearing limestone or other formations with acid for the purpose of increasing production. Hydrochloric or other acid is injected into the formation under pressure. The acid etches the rock, enlarging the pore spaces and passages through which the reservoir fluids flow. The acid is held under pressure for a period of time and then pumped out, after which the well is swabbed and put back into production. Chemical inhibitors combined with the acid prevent corrosion of the pipe. Sandstone reservoirs are often treated with small amounts of acid with other chemicals to decrease damage caused by mud and mud filtrate while drilling.

acid wash *n*: an acid treatment in which an acid mixture is circulated through a wellbore to clean the perforations.

acoustic log *n*: a record of the measurement of porosity done by comparing depth to the time it takes for a sonic impulse to travel through a given length of formation. The rate of travel of the sound wave through a rock depends on the composition of the formation and the fluids it contains. Because other logs can ascertain the type of formation, and because sonic transit time varies with relative amounts of rock and fluid, porosity can usually be determined in this way.

annular pressure *n*: fluid pressure in an annular space, as around tubing within casing.

annular production *n*: production of formation fluids through the production casing annulus.

annular space *n*: 1. the space that surrounds a cylindrical object within a cylinder. 2. the space around a pipe in a wellbore, the outer wall of which may be the wall of either the borehole or the casing; sometimes termed the annulus.

annulus *n*: see *annular space*.

API gravity *n*: the measure of the density or gravity of liquid petroleum products on the North American continent; derived from relative density in accordance with the following equation:

$$\text{API gravity at } 60°\text{F} = [141.5 \div \text{relative density } 60/60°\text{F}] - 131.5$$

API gravity is expressed in degrees, 10° API being equivalent to 1.0, the specific gravity of water. See *gravity*.

artificial lift *n*: any method used to raise oil to the surface through a well after reservoir pressure has declined to the point at which the well no longer produces by means of natural energy. Sucker rod pumps, gas lift, hydraulic pumps, and submersible electric pumps are the most common forms of artificial lift. See *plunger lift*.

B

balloon *v*: to swell or puff up. In reference to tubing under the effects of temperature changes, sucker rod pumping, or high internal pressure, to increase in diameter while decreasing in length. Compare *reverse-balloon*.

barrel *n*: a measure of volume for petroleum products in the United States. One barrel is the equivalent of 42 U.S. gallons or 0.15899 cubic metres (9,702 cubic inches). One cubic metre equals 6.2897 barrels.

battery *n*: 1. an installation of identical or nearly identical pieces of equipment (such as a tank battery or a battery of meters). 2. an electricity storage device.

bbl *abbr*: barrel.

bbl/d *abbr*: barrels per day.

beam pumping unit *n*: a machine designed specifically for sucker rod pumping, using a horizontal member (walking beam) that is worked up and down by a rotating crank to produce reciprocating motion.

blast joint *n*: a tubing sub made of abrasion-resistant material, used in a tubing string where high-velocity flow through perforations or entrained solids may cause external erosion.

blowout preventer *n*: one of several valves installed at the wellhead to prevent the escape of fluids and pressure either in the annular space between the casing and drill pipe or in open hole (i.e., hole with no drill pipe) during drilling and completion operations. Blowout preventers on land rigs are located beneath the rig at the land's surface; on jackup or platform rigs, at the water's surface; and on floating offshore rigs, on the seafloor.

bomb *n*: a thick-walled container, usually steel, used to hold devices that determine and record pressure or temperature in a wellbore. See *bottomhole pressure*.

BOP *abbr*: blowout preventer.

borehole *n*: the wellbore; the hole made by drilling or boring. See *wellbore*.

bottomhole pressure *n*: 1. the pressure at the bottom of a wellbore. It is caused by the hydrostatic pressure of the wellbore fluid and, sometimes, any back-pressure held at the surface, as when the well is shut in with blowout preventers. When mud is being circulated, bottomhole pressure is the hydrostatic pressure plus the remaining circulating pressure required to move the mud up the annulus. 2. the pressure in a well at a point opposite the producing formation, as recorded by a bottomhole pressure bomb. 3. reservoir pressure. See *bottomhole pressure bomb*.

bottomhole pressure bomb *n*: a bomb used to record the pressure in a well at a point opposite the producing formation. See *bomb*.

bottomhole pressure gauge *n*: a gauge to measure bottomhole pressure. See *bottomhole pressure*.

bring in a well *v*: to complete a well and put into producing status.

bring on a well *v*: to bring a well on-line; that is, to put it into producing status.

bubble bucket *n*: a bucket of water into which air from the drill stem is displaced through a hose during the first flow period of a drill stem test. In drill stem testing, the flow into the bubble bucket is an easy way to judge flow and shut-in periods.

burst strength *n*: a pipe or vessel's ability to withstand rupture from internal pressure.

butane *n*: a hydrocarbon molecule consisting of carbon and hydrogen (C_4H_{10}).

C

carbonate rock *n*: a sedimentary rock composed primarily of calcium carbonate (limestone) or calcium magnesium carbonate (dolomite); sometimes makes up petroleum reservoirs.

carbon dioxide *n*: a colorless, odorless gaseous compound of carbon and oxygen (CO_2).

cased hole *n*: a wellbore in which casing has been run. See *casing*.

casing *n*: steel pipe placed in an oil or gas well as drilling progresses to prevent the wall of the hole from caving in during drilling, to prevent seepage of fluids, to provide subsurface control of oil and gas reservoirs, and to provide a means of extracting petroleum if the well is productive. After completion casing is cemented in place to secure the wellbore and provide housing for the tubing. See *tubing*

casing hanger *n*: an arrangement of slips and packing rings, used to suspend casing from a casinghead.

casinghead *n*: a heavy, usually flanged (occasionally screwed or welded) steel fitting connected to the first string of casing; it provides a housing for slips and packing assemblies, allows suspension of intermediate and production strings of casing, and supplies the means for the annulus to be sealed off. Also called a spool in the flanged variety.

centimetre *n*: a unit of length in the SI metric system equal to 0.01 metre (10^{-2} metre). Its symbol is cm.

check valve *n*: a valve that permits flow in one direction only. Commonly referred to as a one-way valve. If the gas or liquid starts to reverse, the valve automatically closes, preventing reverse movement.

choke manifold *n*: an arrangement of piping and special valves, called chokes. In drilling, mud is circulated through a choke manifold when the blowout preventers are closed; a choke manifold is also used to control the pressures encountered during a kick. In well testing, a choke manifold attached to the wellhead allows flow and pressure control for test components downstream.

Christmas tree *n*: the control valves, pressure gauges, and chokes assembled at the top of a well to control the flow of oil and gas after the well has been drilled and completed.

clastic rocks *n pl*: sedimentary rocks composed of fragments of preexisting rocks. Sandstone is a clastic rock.

coal bed methane *n*: natural gas, primarily methane, that occurs naturally in the fractures and matrix of coal beds.

collapse strength *n*: the maximum bearable external pressure that a joint of casing or other tubular can withstand without collapsing.

complete a well *v*: to finish work on a well and bring it to productive status. See *well completion*.

completion fluid *n*: a low-solids fluid or drilling mud used when a well is being completed. It is selected not only for its ability to control formation pressure, but also for the properties that minimize formation damage.

compression packer *n*: a packer that is kept in its set position by pipe weight set down on it.

conductor line *n*: a small-diameter conductive line used in electric wireline operations, such as electric well logging and perforating, in which the transmission of electrical current is required. Compare *wireline*.

connection *n*: 1. a section of pipe or fitting used to join pipe to pipe or pipe to a vessel. 2. a place in electrical circuits where wires join together.

contact *n*: in a petroleum reservoir, a horizontal boundary where different types of fluids meet and mix slightly; for example, a gas-oil or oil-water contact. Also called an interface.

core *n*: a cylindrical sample taken from a formation for geological analysis. Usually a conventional core barrel is substituted for the bit and captures a sample as it penetrates the formation. *v*: to obtain a formation sample for analysis. Samples may also be recovered by use of a wireline tool.

core analysis *n*: laboratory analysis of a core sample to determine porosity, permeability, lithology, fluid content, electrical and radioactive properties, angle of dip, geological age, and probable productivity of the formation.

core barrel *n*: a tubular device, usually from 10 to 60 feet (3 to 18 metres) long, run at the bottom of the drill pipe in place of a bit and used to cut a core sample.

corrodent *n*: a corrosion agent. For example, salt water (oilfield brine) is a corrodent.

corrosion *n*: any of a variety of complex chemical or electrochemical processes by which metal is destroyed through reaction with its environment. Rust (iron oxide: Fe_2O_3) is a product of corrosion.

D

darcy *n*: a unit of measure of permeability. A porous medium has a permeability of 1 darcy when a pressure of 1 atmosphere on a sample 1 centimetre long and 1 square centimetre in cross section will force a liquid of 1 centipoise of viscosity through the sample at the rate of 1 cubic centimetre per second. The permeability of reservoir rocks is usually so low that it is measured in millidarcys.

density log *n*: a special radioactivity log for open-hole surveying that responds to variations in the specific gravity of formations. It is a contact log (i.e., the logging tool is held against the wall of the hole). It emits neutrons and then measures the secondary gamma radiation that is scattered back to the detector in the instrument. The density log is an excellent porosity-measure device, especially for shaley sands. Some trade names are Formation Density Log, Gamma-Gamma Density Log, and Densilog.

differential pressure *n*: the difference between two fluid pressures; for example, the difference between the pressure in a reservoir and in a wellbore drilled in the reservoir, or between atmospheric pressure at sea level and at 10,000 feet (3,048 metres). The difference between the reservoir pressure at flowing and static conditions in a well. The pressure drop across a gas metering device such as an orifice well tester, a pipeline sales meter, or a wellhead choke. Also called pressure differential.

dolomite *n*: a type of sedimentary rock similar to limestone but rich in magnesium carbonate; sometimes a reservoir rock for petroleum.

drag blocks *n pl*: spring-loaded buttons on a packer that provide friction with casing to retard movement of one section of a packer while another section rotates for setting.

drag springs *n pl*: spring-loaded curved metal bands on a packer that serve similarly to drag blocks. See *drag blocks*.

driller's log *n*: a record that describes each formation encountered and lists the drilling time relative to depth, usually in 5- to 10-foot (1.5- to 3-metre) intervals.

drill stem test *n*: the conventional method of formation testing. The basic drill stem test tool consists of a packer or packers, valves or ports that may be opened and closed from the surface, and two or more pressure-recording devices. The tool is lowered on the drill string to the zone to be tested. The packer or packers are set to isolate the zone from the drilling fluid column. The valves or ports are then opened to allow for formation flow while the recorders chart flow pressures, and are then closed, to shut in the formation while the recorders chart static pressures. A sampling chamber traps clean formation fluids at the end of the test. Analysis of the pressure charts is an important part of formation testing.

DST *abbr*: drill stem test.

DST tool *n*: drill stem test tool; used for formation evaluation.

E

energized systems *n pl*: adds nitrogen, carbon dioxide, or both, to water-based and oil-based systems to provide for more rapid cleanup of frac fluid and occasionally to improve sand-carrying characteristics.

F

flange *n*: a projecting rim or edge (as on pipe fittings and openings in pumps and vessels), usually drilled with holes to allow bolting to other flanged fittings.

flow coupling *n*: a tubing sub made of abrasion-resistant material, used in a tubing string where turbulent flow or entrained solids may cause internal erosion.

flowing pressure *n*: pressure registered at the wellhead of a flowing well.

flow-line manifold *n*: in surface well testing, a pipe sub in the flow line that provides points to attach temperature and pressure gauges, outlets for fluid sampling, and inlets for chemical injection during the test.

flow period *n*: in formation testing, an interval of time during which a well is allowed to flow while flow characteristics are being measured.

flow pressure *n*: see *flowing pressure*.

fluid *n*: a substance that flows and yields to any force tending to change its shape. Liquids and gases are fluids.

formation fluid *n*: fluid (such as gas, oil, or water) that exists in a subsurface rock formation.

formation testing *n*: the gathering of pressure data and fluid samples from a formation to determine its production potential before choosing a completion method. Formation testing tools include wireline formation testers, drill stem test tools, and surface test packages.

fossil *n*: a remnant or trace of an animal or plant from the geologic past.

fracture *n*: a crack or crevice in a formation, either natural or induced.

G

gamma ray log *n*: a type of radioactivity well log that records natural radioactivity around the wellbore. Shales generally produce higher levels of gamma radiation and can be detected and studied with the gamma ray tool. In holes where salty drilling fluids are used, electric logging tools are less effective than gamma ray tools.

gas chromatograph *n*: a device used to separate and identify gas compounds by their adhesion to different layers of a filtering medium such as clay or paper, sometimes indicated by color changes in the medium.

gas lift *n*: the process of raising or lifting fluid from a well by injecting gas down the well through a parallel tubing string or through the tubing-casing annulus. Injected gas aerates the fluid to make it exert less pressure than the formation does; consequently, the higher formation pressure forces the fluid out of the wellbore. Gas may be injected continuously or intermittently, depending on the producing characteristics of the well and arrangement of the gas-lift equipment.

gas-lift mandrel *n*: a device installed in the tubing string of a gas-lift well onto which or into which a gas-lift valve is fitted. There are two common types of mandrels. In the conventional gas-lift mandrel, the gas-lift valve is installed as the tubing is placed in the well. Thus, to replace or repair the valve, the tubing string must be pulled. In the side-pocket mandrel, however, the valve is installed and removed by wireline while the mandrel is still in the well, eliminating the need to pull the tubing to repair or replace the valve.

gas-lift valve *n*: a device installed in a gas-lift mandrel. Tubing and casing pressures cause the valve to open and close, thus allowing gas to be injected into the fluid in the tubing to cause the fluid to rise to the surface. See *gas-lift mandrel.*

gathering system *n*: the pipelines and other equipment needed to transport oil, gas, or both from wells to a central point—the gathering station—where there is the accessory equipment required to deliver a clean and salable product to the market or to another pipeline. An oil gathering system includes oil and gas separators, emulsion treaters, gathering tanks, and similar equipment. A gas gathering system includes regulators, compressors, dehydrators, and associated equipment.

gravel *n*: in gravel packing, sand or glass beads of uniform size and roundness.

gravel-pack *v*: to place a slotted or perforated liner in a well and surround it with gravel. See *gravel packing.*

gravel packing *n*: a method of well completion in which a slotted or perforated liner, often wire-wrapped, is placed in the well and surrounded by gravel. If open hole, the well is sometimes enlarged by underreaming at the point where the gravel is packed. The mass of gravel excludes sand from the wellbore but allows continued production.

H

heading *n*: an intermittent flow of oil from a well.

heater *n*: a container or vessel enclosing an arrangement of tubes and a firebox in which 1) an emulsion is heated before further treating or 2) natural gas is heated to prevent hydrate formation.

hold-down button *n*: a hydraulic, toothed device in a packer that uses differential pressure across the packer to grip casing and prevent upward packer movement.

hydrate *n*: an unstable compound consisting of hydrocarbon and water that is formed under reduced temperature and pressure in gathering, compression, and transmission facilities for gas. Hydrates often accumulate in troublesome amounts and impede fluid flow. They resemble snow or ice and decompose at atmospheric pressure.

hydraulic jet perforating *n*: the use of sand pumped through horizontally positioned orifices installed in the tubing string to jet holes in the casing and cement sheath into the formation.

hydraulic-set packer *n*: packer that is set using pressure applied either by pistons actuated by formation pressure or using externally applied pump pressure.

hydrocarbons *n pl*: organic compounds of hydrogen and carbon whose densities, boiling points, and freezing points increase as their molecular weights increase. Although composed of only two elements, hydrocarbons exist in a variety of compounds because of the strong affinity of the carbon atom for other atoms and for itself. The smallest molecules of hydrocarbons are gaseous; the largest are solids. Petroleum is a mixture of many different hydrocarbons.

hydrogen sulfide *n*: a flammable, colorless gaseous compound of hydrogen and sulfur (H_2S) with the odor of rotten eggs. Commonly found in petroleum, it causes the foul smell of petroleum fractions. It is extremely corrosive and poisonous, causing damage to skin, eyes, breathing passages, and lungs, and attacking and paralyzing the nervous system, particularly that part controlling the lungs and heart. Also called hepatic gas or sulfurated hydrogen.

hydrostatic pressure *n*: the force exerted by a body of fluid at rest; it increases directly with the density and the depth of the fluid and is expressed in pounds per square inch or kilopascals. The hydrostatic pressure of fresh water is 0.433 pounds per square inch per foot (9.792 kilopascals per metre) of depth. In drilling, the term refers to the pressure exerted by the drilling fluid in the wellbore. In a water drive field, the term refers to the pressure that may furnish the primary energy for production.

I

igneous rock *n*: a rock mass formed by the solidification of material poured (when molten) into the earth's crust or onto its surface. Granite is an igneous rock.

impermeable *adj*: preventing the passage of fluid. A formation may be porous yet impermeable if there is an absence of connecting passages between the voids within it. See *permeability*.

induction log *n*: see *induction survey*.

induction survey *n*: an electric well log in which the conductivity of the formation rather than the resistivity is measured. Because oil-bearing formations conduct less electricity than water-bearing formations, an induction survey, when compared with resistivity readings, can aid in determining oil and water zones.

inflatable straddle tool *n*: consists of two packers spaced exactly as the straddle-packer DST. It is usually set by pump pressure instead of mechanically set.

J

junk *n*: metal debris lost in a hole. Junk may be a lost bit, pieces of a bit, milled pieces of pipe, wrenches, or any relatively small object that impedes drilling or completion and must be fished out of the hole.

junk basket *n*: a device made up on the bottom of the drill stem or on wireline to catch pieces of junk from the bottom of the hole. Circulating the mud or reeling in the wireline forces the junk into a barrel in the tool, where it is caught and held. When the basket is brought back to the surface, the junk is removed. Also called a junk sub or junk catcher.

K

kill *v*: 1. in drilling, to prevent a threatened blowout by taking suitable pre-ventive measures (e.g., to shut in the well with the blowout preventers, circulate the kick out, and increase the weight of the drilling mud). 2. in production, to stop a well from producing oil and gas so that reconditioning of the well can proceed. Production is stopped by circulating a killing fluid into the hole.

kilopascal (kPa) *n*: 1,000 pascals. See *pascal*.

kPa *abbr*: kilopascal.

L

land a wellhead *v*: to attach casingheads and other wellhead equipment not already in place at the time of well completion.

lateral focus log *n*: a resistivity log taken with a sonde that focuses an electrical current laterally, away from the wellbore, and into the formation being logged. Allows more precise measurement than was possible with earlier sondes. The lateral focus log is a useful means of distinguishing thin rock layers. Also called Laterolog™.

limestone *n*: a sedimentary rock rich in calcium carbonate; sometimes serves as a reservoir rock for petroleum.

loading *n*: occurs when the amount of water or condensate production is too great to be lifted by the velocity of the gas.

M

m *abbr*: metre.

manifold *n*: 1. an accessory system of piping to a main piping system (or another conductor) that serves to divide a flow into several parts, to combine several flows into one, or to reroute a flow to any one of several possible destinations. 2. a pipe fitting with several side outlets to connect it with other pipes. 3. a fitting on an internal-combustion engine made to receive exhaust gases from several cylinders.

marker bed *n*: a distinctive, easily identified rock stratum, especially one used as a guide for drilling or correlation of logs.

mass spectrometer *n*: a device that measures the specific gravity or relative density of a gas by means of the separation of gaseous ions according to their differing mass and charge.

master valve *n*: 1. a large valve located on the Christmas tree and used to control the flow of oil and gas from a well. Also called a master gate. 2. the blind or blank rams of a blowout preventer.

matrix *n*: in rock, the fine-grained material between larger grains and in which the larger grains are embedded. A rock matrix may be composed of fine sediments, crystals, clay, or other substances. Solid portions of a porous rock that does not include pores.

md *abbr*: millidarcy.

metamorphic rock *n*: a rock derived from preexisting rocks by mineralogical, chemical, and structural alterations caused by processes within the earth's crust. Marble is a metamorphic rock.

methane *n*: a light, gaseous, flammable paraffinic hydrocarbon, CH_4, that has a boiling point of –25°F (and is the chief component of natural gas and an important basic hydrocarbon for petrochemical manufacture).

metre (m) *n*: the fundamental unit of length in the international system of measurement (SI). It is equal to about 3.28 feet, 39.37 inches, or 100 centimetres.

microresistivity log *n*: a resistivity logging tool consisting of a spring device and a pad. While the spring device holds the pad firmly against the borehole sidewall, electrodes in the pad measure resistivities in mud cake and nearby formation rock.

millidarcy *n*: one-thousandth of a darcy.

millivolt *n*: one-thousandth of a volt.

miniaturized completion *n*: a well completion in which the production casing is less than 4.5 inches (11.43 centimetres) in diameter.

mud cake *n*: the sheath of mud solids that forms on the wall of the hole when liquid from mud filters into the formation. Also called wall cake or filter cake.

mud filtrate *n*: the liquid which is expelled from drilling mud during the formation of mud cake.

mud log *n*: a record of information derived from examination of drilling fluid and drill bit cuttings. See *mud logging*.

mud logging *n*: the recording of information derived from examination and analysis of formation cuttings made by the bit and of mud circulated out of the hole. A portion of the mud is diverted through a gas-detecting device. Cuttings brought up by the mud are examined under ultraviolet light to detect the presence of oil or gas. Mud logging is often carried out in a portable laboratory set up at the well.

multiple completion *n*: an arrangement for producing a well in which one wellbore penetrates two or more petroleum-bearing formations. In one type, multiple tubing strings are suspended side by side in the production casing string, each a different length and each packed off to prevent the commingling of different reservoir fluids. Each reservoir is then produced through its own tubing string. Alternatively, a small-diameter production casing string may be provided for each reservoir, as in multiple slim hole or multiple tubingless completion.

N

natural gas *n*: a naturally occurring hydrocarbon consisting largely of methane. Gas exists in the gaseous phase or dissolved in oil. See *hydrocarbons*, *methane*.

neutral packer *n*: a packer that, unlike compression or tension packers, requires neither set-down weight nor upstrain pull to remain set.

neutron log *n*: a radioactivity well log used to determine formation porosity. The logging tool bombards the formation with neutrons. When the neutrons strike hydrogen atoms in water or oil, gamma rays are released. Since water or oil exists only in pore spaces, a measurement of the gamma rays indicates formation porosity.

nozzle *n*: 1. a passageway through jet bits that allows the drilling fluid to reach the bottom of the hole and flush the cuttings through the annulus. Nozzles come in different sizes that can be interchanged on the bit to allow more or less flow. 2. the part of the fuel system of an engine that has small holes in it to permit fuel to enter the cylinder. Properly known as a fuel-injection nozzle, but also called a spray valve. The needle valve is directly above the nozzle.

O

oil and gas separator *n*: an item of production equipment used to separate liquid components of the well stream from gaseous elements. Separators are either vertical or horizontal and either cylindrical or spherical in shape. Separation is accomplished principally by gravity, the heavier liquids falling to the bottom and gas rising to the top. A float valve or other liquid-level control regulates the level of oil in the bottom of the separator.

open hole *n*: 1. any wellbore in which casing has not been set. 2. open or cased hole in which no drill pipe or tubing is suspended. 3. the portion of the wellbore that has no casing.

open-hole completion *n*: a method of preparing a well for production in which no production casing or liner is set opposite the producing formation. Reservoir fluids flow unrestricted into the open wellbore. An open-hole completion has limited use in rather special situations. Also called a barefoot completion.

oriented perforating *n*: a perforating technique that uses sensing instruments in a perforating gun to make perforations in a specific direction, often used in completions involving multiple production casing strings to perforate one string without damaging another. Also called directional perforating.

orifice plate *n*: a sheet of metal, usually circular, in which a hole of specific size is made for use in an orifice fitting.

P

packer *n*: a piece of downhole equipment, consisting of a sealing device, a holding or setting device, and an inside passage for fluids, used to block the flow of fluids through the annular space between pipe and the wall of the wellbore by sealing off the space between them. In production, it is usually made up in the tubing string some distance above the producing zone. A packing element expands to prevent fluid flow except through the packer and tubing. Packers are classified according to configuration, use, and method of setting and whether or not they are retrievable (that is, whether they can be removed when necessary, or whether they must be milled or drilled out and thus destroyed).

packer fluid *n*: a liquid, usually salt water or oil, but sometimes mud, used in a well when a packer is between the tubing and the casing. Packer fluid must be heavy enough to shut off the pressure of the formation being produced, must not stiffen or settle out of suspension over long periods of time, and must be noncorrosive.

packing elements *n pl*: the set of washer-shaped pieces, made of dense rubber and other materials, encircling a packer. Packer elements are designed to expand against casing or formation face to seal off the annulus.

paraffin *n*: a hydrocarbon having the formula C_nH_{2n+2} (e.g., methane, CH_4; ethane, C_2H_6). Heavier paraffin hydrocarbons (i.e., $C_{18}H_{38}$ and heavier) form a waxlike substance that is called paraffin. These heavier paraffins often accumulate on the walls of tubing and other production equipment, restricting or stopping the flow of desirable lighter paraffins.

parallel strings *n pl*: in multiple completion, the arrangement of a separate tubing string for each zone produced, usually with all zones isolated by packers.

pascal *n*: the accepted metric unit of measurement for pressure and stress and a component in the measurement of viscosity. A pascal is equal to a force of 1 newton acting on an area of 1 square metre. Its symbol is Pa.

pay sand *n*: the producing formation, one that may not be sandstone. It is also called pay, pay zone, and producing zone.

perforate *v*: to pierce the casing wall and cement to provide holes through which formation fluids may enter or to provide holes in the casing so that materials may be introduced into the annulus between the casing and wall of the borehole. Perforating is accomplished by lowering into the well a perforating gun, or perforator, that fires electrically detonated bullets or shaped charges.

perforated completion *n*: 1. a well completion method in which the producing zone or zones are cased through, cemented, and perforated to allow fluid flow into the wellbore. 2. a well completed by this method.

perforation *n*: a hole made in the casing, cement, and formation, through which formation fluids enter a wellbore. Usually several perforations are made at a time.

permanent packer *n*: a nonretrievable type of packer that must be drilled or milled out for removal. Most permanent packers are seal-bore packers or wireline-set packers.

permeability *n*: 1. a measure of the ease with which fluids can flow through a porous rock. 2. the fluid conductivity of a porous medium. 3. the ability of a fluid to flow within the interconnected pore network of a porous medium. See *relative permeability*.

permeability barrier *n*: a hindrance to movement of fluids within the formation rock of a reservoir. Permeability barriers include obvious problems such as shale lenses and calcite or clay deposits, and less obvious ones, such as porosity changes.

permeable *adj*: allowing the passage of fluid. See *permeability*.

petroleum *n*: a mixture of hydrocarbons. See *hydrocarbons*.

photoelectric effect (Pe) *n*: the absorption of gamma rays that results in ejection of electrons from an atom. Photoelectric effect occurs when light of sufficient energy falls on an atom and causes it to lose electrons. The energy of the light actually tears electrons away from the atoms of a substance.

pinch-out *n*: a geological structure that forms a trap for oil and gas when a porous and permeable rock ends at or stops against an impervious formation.

plunger lift *n*: a system employing a free falling plunger, which is an open-while-falling, closed-while-rising valve. This valve provides a seal between incoming wellbore fluids and accumulated liquids in the tubing to improve unloading of liquids. The system provides alternating flowing and shut-in times determined by a surface controller actuated either by a time clock or well pressure.

porosity *n*: 1. the condition of being porous (such as a rock formation). 2. the ratio of the volume of empty space to the volume of solid rock in a formation, indicating how much fluid a rock holds.

precipitate *n*: a substance, usually a solid, that separates from a fluid because of a chemical or physical change in the fluid. *v*: to separate in this manner.

pressure *n*: the force that a fluid (liquid or gas) exerts uniformly in all directions within a vessel, pipe, hole in the ground, and so forth, such as that exerted against the inner wall of a tank or that exerted on the bottom of the wellbore by a fluid. Pressure is expressed in terms of force exerted per unit of area, as pounds per square inch (psi), grams or kilograms per square centimetre, or kilopascals.

pressure maintenance *n*: a method of increasing ultimate oil recovery by injecting gas, water, or other fluids into the reservoir to reduce or eliminate a decline in pressure before reservoir pressure has dropped appreciably, usually early in the life of the field.

pressure management *n*: the efforts made to maintain reservoir drive at efficient levels for maximum recovery of oil and gas. Pressure management involves controlled production rates based on the daily rate of withdrawal and monthly production allowed for each well and takes into account any pressure maintenance methods in use.

primary porosity *n*: natural porosity in petroleum reservoir sand or rocks, i.e., the porosity developed during the original sedimentation process. Formations having this property (such as sand) are usually granular.

production *n*: 1. the phase of the petroleum industry that deals with bringing the well fluids to the surface and separating them and with storing, gauging, and otherwise preparing the product for the pipeline. 2. the amount of oil or gas produced in a given period.

production casing *n*: the last string of casing liner that is set in a well, inside of which is usually suspended the tubing string.

production maintenance *n*: the efforts made to minimize the decline in a well's production. Production maintenance includes practices such as acid-washing of casing perforations to dissolve mineral deposits, scraping or chemical injection to prevent paraffin buildup, and various measures taken to control corrosion and erosion damage.

progressing cavity pump *n*: a type of subsurface pump in which the rods are rotated instead of reciprocated. The pump consists of an auger-like assembly that rotates inside a cylinder. The reverse screw action forces fluid up through the tubing to the subsurface. Especially useful in reservoirs containing heavy crude oil or crude oil with a high-solids content.

propane *n*: a hydrocarbon molecule consisting of carbon and hydrogen (C_3H_8).

Q

quartz *n*: a hard mineral composed of silicon dioxide (silica), a common component in igneous, metamorphic, and sedimentary rocks.

R

recompletion *n*: the changing of producing interval at a time subsequent to the original completion.

relative permeability *n*: the permeability of a reservoir to oil, gas, or water relative to absolute permeability. Example: $K_{ro} = K_o/K$ where K_{ro} is defined as oil relative permeability. K_o is defined as oil permeability, and K is defined as absolute permeability. Similar relationships are true for gas and water. Consequently, in a reservoir where oil, gas, and water usually coexist, the absolute permeability reflects the productivity of the total rock pore environment. Relative permeability defines the ability of the rock to transmit a particular fluid. Relative permeability varies with various rock properties including saturation and wettability. Thus, productivity of oil, gas, and water vary with time as saturations change.

remedial stimulation *n*: the restoration of a well's performance using chemicals such as acid and paraffin solvents to dissolve restricting materials in the perforations and near the wellbore.

reservoir *n*: a subsurface, porous, permeable or naturally fractured rock body in which oil or gas is stored. Most reservoir rocks are limestones, dolomites, sandstones, or a combination of these. The four basic types of hydrocarbon reservoirs are oil, volatile oil, dry gas, and gas condensate.

An oil reservoir generally contains three fluids—gas, oil, and water—with oil the dominant product. In the typical oil reservoir, these fluids become vertically segregated because of their different densities. Gas, the lightest, occupies the upper part of the reservoir rocks; water, the lower part; and oil, the intermediate section. In addition to its occurrence as a cap or in solution, gas may accumulate independently of the oil; if so, the reservoir is called a gas reservoir. Associated with the gas, in most instances, are salt water and some oil.

Volatile oil reservoirs are exceptional in that during early production they are mostly productive of light oil plus gas, but, as depletion occurs, production can become almost totally completely gas. Volatile oils are usually good candidates for pressure maintenance projects, which can result in increased reserves.

In the typical dry gas reservoir natural gas exists only as a gas and production is only gas plus fresh water that condenses from the flow stream reservoir.

In a gas condensate reservoir, the hydrocarbons may exist as a gas, but, when brought to the surface, some of the heavier hydrocarbons condense and become a liquid.

reverse-balloon *v*: in reference to tubing under the effects of temperature changes, sucker rod pumping, or high external pressure, to decrease in diameter while increasing in length. Compare *balloon*.

routine well service *n*: routine repairs including but not limited to sucker rod failures, tubing leaks, subsurface pump repairs, and attendant equipment and material repairs and replacement.

run in *v*: to go into the hole with tubing, drill pipe, tools, and so forth.

S

sanding off *n*: a phenomenon that occurs when sand and sediment settle out of oil being produced from a well and impedes or blocks the flow of produced fluids from the well.

sandstone *n*: a sedimentary rock composed of individual mineral grains of rock fragments between 0.06 and 2 millimetres (0.002 and 0.079 inches) in diameter and cemented together by silica, calcite, iron oxide, and so forth. Sandstone is commonly porous and permeable and therefore a type of rock in which to find a petroleum reservoir.

S&W *abbr*: sediment and water.

saturation *n*: a state of being filled or permeated to capacity. Sometimes used to mean the degree or percentage of saturation (as, the saturation of the pore space in a formation or the saturation of gas in a liquid, both in reality meaning the extent of saturation). See *relative permeability*.

screen liner *n*: a pipe that is perforated and often arranged with a wire wrapping to act as a sieve to prevent or minimize the entry of sand particles into the wellbore. Also called a screen pipe.

seal bore *n*: a smooth, polished bore in pipe or in a packer, designed to provide an inner surface against which to make an effective seal.

seal-bore packer *n*: a packer containing a smooth polished bore to receive a tubing seal assembly.

seating nipple *n*: a special tube installed in a string of tubing, having machined contours to fit a matching wireline tool with locking pawls. It is used to hold a regulator, choke, or safety valve; to anchor a pump; or to permit installation of gas-lift valves. Also called a landing nipple. A smooth bore tubing nipple, having an inside diameter less than tubing, used to accept the seal assembly of a rod pump, as a stop for a plunger, as a safety stop for tubing swabs.

secondary porosity *n*: porosity created in a formation after it has formed, because of dissolution or stress distortion taking place naturally, or because of treatment by acid or injection of coarse sand.

secondary recovery *n*: the injecting of fluids, most frequently water, into the reservoir in order to restore reservoir pressure and to increase the amount of hydrocarbons ultimately produced.

sediment and water (S&W) *n*: the water and other extraneous material present in crude oil. Usually, the S&W content must be quite low before a pipeline will accept the oil for delivery to a refinery. The amount acceptable depends on a number of factors but usually lies between less than 5 percent to a small fraction of 1 percent.

sedimentary rock *n*: a rock composed of materials that were transported to their present position by wind or water. Sandstone, shale, and limestone are sedimentary rocks.

seismic survey *n*: a survey conducted using explosives or other means of propagating a sound wave from the surface down through the rock deposits. As the sound passes through succeeding rock layers, faults, discontinuities, and different reservoir fluids, the reflections are recorded by geophones at the surface that can lead to helpful interpretations for locating oil and gas accumulations.

separator *n*: a cylindrical or spherical vessel used to isolate the components in streams of mixed fluids. See *oil and gas separator*.

shale *n*: a fine-grained sedimentary rock composed of consolidated silt and clay or mud. Shale is the most frequently occurring sedimentary rock.

show *n*: the appearance on the surface of oil or gas in drilling fluids, in cuttings, samples, or cores from a drilling well.

shut-in period *n*: an interval of time during which a well is shut in to allow pressure buildup while pressure behavior is being measured during a formation test.

single-string dual completion *n*: 1. a method of well completion in which two pay zones are produced by directing fluids from one zone through a tubing string and from the other zone through the annulus. 2. the tool assembly used to complete a well by this method.

sinker bar *n*: a heavy weight or bar placed on or near a lightweight wireline tool. The bar provides weight so that the tool will lower properly into the well.

slick line *n*: see *wireline*.

sliding sleeve *n*: a slotted device placed in a string of tubing and operated by a wireline tool (1) to open or close orifices, thus permitting circulation between the tubing and the annulus, or (2) to open or shut off production from alternate intervals in a well; also called a sliding side door or circulating sleeve.

slips *n pl*: wedge-shaped pieces of metal with teeth or other gripping elements that are used to prevent pipe from slipping down into the hole or to hold pipe in place. Rotary slips fit around the drill pipe and wedge against the master bushing to support the pipe. Power slips are pneumatically or hydraulically actuated devices that allow the crew to dispense with the manual handling of slips when making a connection. Packers and other downhole equipment are secured in position by slips that engage the inner surface of casing.

sonde *n*: a logging tool assembly, especially the device in the logging assembly that senses, gathers, and transmits formation data.

specific gravity *n*: the ratio of the weight of a given volume of a substance at a given temperature to the weight of an equal volume of a standard substance at the same temperature. For example, if 1 cubic inch of water at 39°F (3.9°C) weighs 1 unit and 1 cubic inch of another solid or liquid at 39°F weighs 0.95 unit, then the specific gravity of the substance is 0.95. In determining the specific gravity of gases, the comparison is made with the standard of air or hydrogen.

static pressure *n*: the pressure exerted by a fluid upon a surface that is at rest in relation to the fluid. The pressure exhibited at the surface or point downhole during the time the well is shut in. Surface or bottomhole pressure after sufficient time has elapsed for the pressure to become stable.

stimulation *n*: the action of attempting to improve and enhance a well's performance by the application of horsepower using pumping equipment, placing sand in artificially created fractures in rock, or using chemicals such as acid to dissolve the soluble portion of the rock.

stripper rubber *n*: 1. a rubber disk surrounding drill pipe or tubing that removes mud as the pipe is brought out of the hole. 2. the pressure-sealing element of a stripper-type blowout preventer.

subsurface safety valve *n*: see *tubing safety valve*.

sucker rod pumping *n*: a method of artificial lift in which a subsurface pump located at or near the bottom of the well and connected to a string of sucker rods is used to lift the well fluid to the surface. The weight of the rod string and fluid is counterbalanced by weights attached to a reciprocating beam or to the crank member of a beam pumping unit or by air pressure in a cylinder attached to the beam.

surface safety valve *n*: a valve, mounted in the Christmas tree assembly, that stops the flow of fluids from the well if damage occurs to the assembly.

surface test tree *n*: in surface well testing, a temporary set of valves installed for flow control on a wellhead with no Christmas tree.

swab off *v*: to pull off during a trip into or out of the hole because of pressure differential. For example, if a packer is run in too quickly, the pressure differential across the packer swabs off the packing elements, making it necessary to trip back out to replace them.

T

temperature *n*: a measure of heat or the absence of heat, expressed in degrees Fahrenheit or Celsius. The latter is the standard used in countries on the metric system.

tensile strength *n*: the greatest longitudinal stress that a metal can bear without tearing apart. A metal's tensile strength is greater than its yield strength.

tension packer *n*: a packer that is held in its set position by upstrain, or upward pull.

thief zone *n*: a low pressure reservoir or zone that steals fluids from a higher pressure reservoir or zone and prevents their production to the surface.

tortuosity *n*: the relative degree of crookedness of a hydrocarbon flow path through a reservoir.

tubing *n*: relatively small-diameter pipe that is run into a well to serve as a conduit for the passage of oil and gas to the surface.

tubing anchor *n*: a device that holds the lower end of a tubing string in place by means of slips, used to prevent tubing movement when no packer is present. Especially for sucker rod pumping applications.

tubing-conveyed perforating *n*: a system in which the tubing is used to convey the perforating guns to the target reservoir and charges are detonated either by dropping a bar or by the application of external pressure.

tubing hanger *n*: an arrangement of slips and packing rings used to suspend tubing from a tubing head.

tubing head *n*: a flanged or threaded fitting that supports the tubing string, seals off pressure between the casing and the outside of the tubing, and provides a connection that supports the Christmas tree.

tubing head swivel *n*: a special pipe sub made up at the upper end of the tubing test string that allows tubing rotation during a well test or remedial operations.

tubingless completion *n*: a method of producing a well in which only production casing is set through the pay zone, with no tubing or inner production string used to bring formation fluids to the surface. This type of completion has limited application to low-pressure, dry-gas reservoirs and to low reserve medium-pressure reservoirs.

tubing movement *n*: changes in tubing length due to changes in pressure, temperature, or drag from sucker rod strokes. Also called breathing.

tubing safety valve *n*: a device installed in the tubing string of a producing well to shut in the flow of production if the flow exceeds a preset rate. Tubing safety valves are widely used in offshore wells to prevent pollution if the wellhead fails for any reason. Also called subsurface safety valve.

tubing seal assembly *n*: a device encircled with rubber rings made up on tubing and designed to fit into a seal-bore packer.

tubing test string *n*: a string of tubing run in a well for a production test in which a well test surface package is used. A tubing test string is similar to a normal tubing string except that it may contain special valves and other tools needed for the well test.

tubular goods *n pl*: any kind of pipe. Oilfield tubular goods include tubing, casing, drill pipe, and line pipe. Also called tubulars.

tubulars *n pl*: shortened form of tubular goods.

V

viscosity *n*: a measure of the resistance of a liquid to flow. The internal friction resulting from the combined effects of cohesion and adhesion brings about resistance. The viscosity of petroleum products is commonly expressed in terms of the time required for a specific volume of the liquid to flow through an orifice of a specific size.

volt (V) *n*: the unit of electric potential, voltage, or electromotive force in the metric system.

vug *n*: a cavity in a rock.

water-base acid *n*: most commonly hydrochloric acid in water, the solution is used to perform massive acid fractures in carbonate reservoirs. Acid concentrations of 22 percent to 28 percent strength provide maximum dissolving of carbonate rock. Small- to medium-sized sand can be added to the system to provide scouring of the fracture faces, along with corrosion inhibitors, deemulsifiers, and clay stabilizing chemicals. Fresh water, in volumes of one to three times the acid volumes, is pumped between the various acid stages to dissolve sludge, which is a result of acid reacting with carbonate.

water-base gel *n*: a solution used in water-base fracturing fluid that gels to protect sensitive sandstone reservoirs.

wellbore *n*: a borehole; the hole drilled by the bit. A wellbore may have casing in it or it may be open (uncased); or a portion of it may be cased, and a portion of it may be open. Also called a borehole or hole.

well completion *n*: 1. the activities and methods of preparing a well for the production of oil and gas; the method by which one or more flow paths for hydrocarbons is established between the reservoir and the surface. 2 the system of tubulars, packers, and other tools installed beneath the wellhead in the production casing; that is, the tool assembly that provides the hydrocarbon flow path or paths.

wellhead *n*: the equipment installed at the surface of the wellbore. A wellhead includes such equipment as the casinghead and tubing head. *adj:* pertaining to the wellhead (e.g., wellhead pressure).

well servicing *n*: relating to well-servicing work, (e.g., a well-servicing company). Repairing or maintaining a well.

well test *n*: a test done on a well to determine its production characteristics.

well test report *n*: the detailed results of a well test that present all test data, the calculations drawn from the data, and conclusions pertaining to the well's production potential.

well test surface package *n*: in well testing, an assembly of equipment that is attached to the wellhead of the well to be tested. The main component of a well test surface package is an oil and gas separator.

wireline *n*: a small diameter metal line used in wireline operations; also called slick line. Compare *conductor line*.

wireline formation tester *n*: a formation fluid sampling device, actually run on conductor line rather than wireline, that also logs flow and shut-in pressure in rock near the borehole. A spring mechanism holds a pad firmly against the sidewall while a piston creates a vacuum in a test chamber. Formation fluids enter the test chamber through a valve in the pad. A recorder logs the rate at which the test chamber is filled. Fluids may also be drawn to fill a sampling chamber. Wireline formation tests may be done any number of times during one trip in the hole, so they are very useful in formation testing.

wireline operations *n pl*: the lowering of mechanical tools, such as valves and fishing tools, into the well for various purposes. Electric wireline operations, such as electric well logging and perforating, involve the use of conductor line, which in the oil patch is commonly but erroneously called wireline.

wireline well logging *n*: the recording of subsurface characteristics by wireline (actually conductor line) tools. Wireline well logs include acoustic logs, caliper logs, radioactivity logs, magnetic resonance logs, and resistivity logs.

workover *n*: the performance of one or more of a variety of remedial operations on a producing oilwell to try to increase production. Examples of workover jobs are deepening, plugging back, pulling and resetting liners, squeeze cementing, and so forth. See *recompletion*.

Y

Y-sub *n*: in multiple packer completion, a device, made up just below the uppermost packer, that switches tubular fluids to the annulus and annular fluids to the tubing above that packer. Also called a Y-block.

Z

zone isolation *n*: the practice of separating producing formations from one another by means of casing, cement, and packers for the purposes of pressure control and maintenance, as well as the prevention of mixing of fluids from separate formations.

Review Questions

LESSONS IN ROTARY DRILLING

Unit II, Lesson 5: Testing and Completing

Multiple Choice

Pick the *best* answer from the choices and place the letter of that answer in the blank provided.

_________ 1. Two types of rock usually hold hydrocarbons. They are—

a. metamorphic and igneous.
b. sedimentary and igneous.
c. clastic and carbonate.
d. granite and limestone.

_________ 2. Forecasters anticipate that about 10 percent of the U.S.'s natural gas production will come from—

a. salt domes.
b. coal bed methane.
c. Alaskan natural gas.
d. none of the above

_________ 3. Porosity is—

a. a measure of how easily fluids flow within a formation.
b. that part of the rock that is solid.
c. not of concern when evaluating oil and gas reservoirs.
d. the volume of pore space divided by bulk volume of solid rock.

_________ 4. Permeability is—

a. a measure of how easily fluids move through rock.
b. the volume of pore space divided by bulk volume of solid rock.
c. not required in a reservoir rock.
d. unrelated to oil and gas production.

_________ 5. Tortuosity is—

a. a measure of the flow path of hydrocarbons through a rock.
b. usually very high in sandstone reservoirs.
c. the volume of pore space divided by bulk volume of solid rock.
d. unrelated to oil and gas production.

_________ 6. Saturation is—

a. a measure of how easily fluids move through a rock.
b. a measure of the percentage of total pore space filled by a fluid.
c. a measure of how much seawater a reservoir holds.
d. immovable water in a reservoir.

_________ 7. Two factors that strongly affect the behavior of reservoir fluids are—
   a. pressure and temperature.
   b. pressure and depth.
   c. temperature and depth.
   d. pressure and fluid.

_________ 8. Which of the following is *not* used to evaluate subsurface formations?
   a. Driller's logs
   b. Core samples
   c. Seismic surveys
   d. Surface indentations

_________ 9. The type of log that records only such items as depth and thickness of each formation bed is a—
   a. mud log.
   b. wireline log.
   c. driller's log.
   d. pine log.

_________ 10. A mud logger checks drilling mud for—
   a. oil.
   b. gas.
   c. cuttings.
   d. all of the above

_________ 11. Two types of cores are—
   a. shallow and deep.
   b. barrel and sidewall.
   c. barrel and lock.
   d. barrel and stock.

_________ 12. The type of log that measures the inner diameter of the wellbore is a—
   a. spontaneous potential log.
   b. resistivity log.
   c. caliper log.
   d. radioactivity log.

_________ 13. One of the most common logs is the—
   a. spontaneous potential log.
   b. resistivity log.
   c. caliper log.
   d. radioactivity log.

_________ 14. The type of log that displays its readings in ohm•meters is the—
a. spontaneous potential log.
b. resistivity log.
c. caliper log.
d. radioactivity log.

_________ 15. A gamma ray log is a—
a. spontaneous potential log.
b. resistivity log.
c. caliper log.
d. radioactivity log.

_________ 16. A type of log that measures induced radiation in a formation rock is a(n)—
a. gamma ray log.
b. neutron log.
c. acoustic log.
d. caliper log.

_________ 17. Formation testing may be done—
a. in open hole.
b. in cased hole.
c. after completion, as a status check.
d. all of the above

_________ 18. A quick, inexpensive way to measure pressures at specific depths is to use a—
a. radioactivity log.
b. drill stem test tool.
c. surface well test.
d. none of the above

True or False

Put a T for *true* and an F for *false* in the blank next to each statement.

_________ 19. A drill stem test (DST) is a kind of temporary and partial completion of a well.

_________ 20. Drill stem tests cannot be run in cased hole.

_________ 21. Two types of drill stem tests are the single-packer DST tool and the upside-down DST tool.

_________ 22. Downhole temperature is one item that a drill stem test tool cannot measure.

________ 23. It is usually not necessary to condition the hole before running a drill stem test tool.

________ 24. A cushion of water or compressed gas, when placed inside the drill stem during a drill stem test, can prevent hydrostatic pressure from crushing the drill stem.

________ 25. Most drill stem tests include two periods of flow, each of which is followed by a shut-in period.

________ 26. In a drill stem test, produced fluids are usually disposed of without being examined.

________ 27. A successful drill stem test can show permeability changes in a formation but cannot detect gas-oil or oil-water contacts.

Fill in the Blanks

Fill in the blanks with the appropriate word or phrase. Pick the correct term from those listed below. (Some terms may be used more than once.)

choke manifold	surface test tree
flow-line manifold	tank
formation testing	tubinghead swivel
heater	well
separator	wellhead
surface safety valves	well test report

Surface well tests are designed to test both specific zones and the entire (28) ________. Surface well tests need only a (29) ________ to attach to. Portable well test equipment can include a (30) ________ ________, a (31) ________ ________ ________, (32) ________ ________ ________, a (33) ________ ________ ________, a (34) ________ ________, a (35) ________, a (36) ________, and a (37) ________. The (38) ________ actually tests the well's production. A detailed (39) ________ ________ ________ is filled out to use in planning the well completion. Generally, (40) ________ ________ tells what fluids will be produced and at what rates they will be produced.

Multiple Choice

Pick the *best* answer from the choices and place the letter of that answer in the blank provided.

_________ 41. The two basic methods of well completion are a(n)—

a. closed hole and liner.
b. open hole and liner.
c. open hole and perforated.
d. multiple and triple.

_________ 42. A coal bed is a good candidate for a(n)—

a. open-hole completion.
b. perforated liner completion.
c. closed-hole completion.
d. horizontal completion.

_________ 43. Most wells use a(n)—

a. open-hole completion.
b. perforated completion.
c. closed-hole completion.
d. horizontal completion.

_________ 44. A company drills a well into a reservoir that is made up of loose, unconsolidated sand. In this case, the company would likely complete the well using a(n)—

a. perforated completion.
b. closed-hole completion.
c. open-hole completion.
d. screen liner and gravel pack.

_________ 45. One reason that tubing is used in most completed wells is—

a. tubing is expensive and difficult to replace.
b. casing is expensive and difficult to replace.
c. both a and b
d. none of the above

_________ 46. A well that has cemented casing run through the production zone, is perforated, and produces through a tubing string is considered to be—

a. a conventional completion.
b. an unconventional completion.
c. a screen liner completion.
d. an open-hole completion.

________ 47. One inexpensive option for completing a shallow, low-pressure well that produces only dry gas is a—

a. tubing completion.
b. tubingless completion.
c. casingless completion.
d. casing and tubing completion.

________ 48. The maximum pull that tubing can bear before failing is the tubing's—

a. burst strength.
b. tensile strength.
c. collapse strength.
d. none of the above

True or False

Put a T for *true* and an F for *false* in the blank next to each statement.

________ 49. A packer is an optional completion tool but it is commonly used.

________ 50. Virtually all packers have packing elements and slips.

________ 51. Some packers are equipped with hydraulic hold-down buttons.

________ 52. If a packer requires rotation to activate its setting mechanisms, it will probably be equipped with drag blocks.

________ 53. Retrieval packers, once set in the well, essentially become a permanent part of the well and must be drilled out, rather than be replaced.

________ 54. Almost always, a seating, or landing, nipple is run on a tubing string.

________ 55. A sliding sleeve is made up in the tubing string to prevent communication between the tubing and the tubing-casing annulus.

________ 56. A check valve usually allows flow in two directions.

________ 57. Subsurface safety valves ensure that production continues to flow although the wellhead may have failed.

________ 58. A blast joint is a thick tubing sub designed to withstand erosion.

________ 59. A flow coupling is installed in a tubing string to prevent the outside of the tubing from becoming worn from erosive elements in the produced fluids.

Multiple Choice

Pick the *best* answer from the choices and place the letter of that answer in the blank provided.

_________ 60. Jet perforating works on the principle of—

a. actual bullets being fired through the casing and cement and into the formation.
b. a mechanical device punching a hole into the casing and cement.
c. propelling a high-energy stream of fine particles through the casing and cement and into the formation.
d. none of the above

_________ 61. Perforating guns include—

a. retrievable types.
b. semi-expendable types.
c. expendable types.
d. all of the above

_________ 62. A completion fluid is usually—

a. a low solids brine.
b. an oil-base fluid.
c. normal drilling mud.
d. a and b

_________ 63. Fluids used in hydraulic fracturing include—

a. water-based gel.
b. water-based acid.
c. fresh water
d. all of the above

Answers to Review Questions

LESSONS IN ROTARY DRILLING

Unit II, Lesson 5: Testing and Completing

1. c
2. b
3. d
4. a
5. a
6. b
7. a
8. d
9. c
10. d
11. b
12. c
13. a
14. b
15. d
16. b
17. d
18. d
19. T
20. F
21. F
22. F
23. F
24. T
25. T
26. F
27. F
28. well
29. wellhead

Answers to 30 through 37 may be given in any order

30. tubinghead swivel
31. surface test tree
32. surface safety valves
33. flow-line manifold
34. choke manifold
35. heater
36. separator
37. tank
38. separator
39. well test report
40. formation testing
41. c
42. a
43. b
44. d
45. b
46. a
47. b
48. b
49. T
50. T
51. T
52. T
53. F
54. T
55. F
56. F
57. F
58. T
59. F
60. c
61. d
62. d
63. d